Solar Thermal Thruster

Minchao Huang · Jianjun Wu · Jian Li ·
Yuqiang Cheng

Solar Thermal Thruster

Minchao Huang
National University of Defense Technology
Changsha, China

Jian Li
National University of Defense Technology
Changsha, China

Jianjun Wu
National University of Defense Technology
Changsha, China

Yuqiang Cheng
National University of Defense Technology
Changsha, China

ISBN 978-981-97-7489-0   ISBN 978-981-97-7490-6 (eBook)
https://doi.org/10.1007/978-981-97-7490-6

Jointly published with National University of Defense Technology Press
The print edition is not for sale in China (Mainland). Customers from China (Mainland) please order the print book from: National University of Defense Technology Press.

This work was supported by National University of Defense Technology Press.

This Springer imprint is published by the registered company Springer Nature Singapore Pte Ltd.
The registered company address is: 152 Beach Road, #21-01/04 Gateway East, Singapore 189721, Singapore

If disposing of this product, please recycle the paper.

Preface

In a solar thermal propulsion (STP) system, concentrators are used to gather sunlight to heat the propellant temperature above 2200 K, Laval nozzle expansion is used to generate thrust, and hydrogen is used as the propellant so that the theoretical specific impulse can reach more than 800 s, achieving a continuous thrust between 0.1 and 1 N. The specific impulse and thrust levels of the STP system range between those of a chemical propulsion system and an electric propulsion system, and the STP system offers great performance advantages in space missions such as orbit transfers. Solar thermal thrusters generally use an indirect heating method of sunlight and use a high-temperature wall surface to heat the propellant; therefore, improving the heat exchange efficiency of the heat exchanger core can further increase the performance parameters of solar thermal thrusters, including thrust and specific impulse. This book provides an integrated design of a solar thermal thruster, effectively combines regenerative cooling methods and heat transfer technology using laminates and achieves a photothermal conversion efficiency of 86% for the thruster, which can maximize the use of the received solar energy. In this book, numerical simulations and experimental studies are performed to assess the radiation and convective heat transfer processes inside solar thermal thrusters, and an optimization analysis of the space application of solar thermal thrusters is performed.

Studies outside China have shown that refractive secondary concentrators (RSCs) of solar thermal thrusters easily crack at high temperatures. This book uses regenerative cooling technology to carry out an integrated design of secondary concentrators and thrust chambers, carries out numerical simulation and experimental studies and uses optical-thermal coupling and fluid-solid coupling methods to conduct a numerical simulation of radiation heat transfer and regenerative cooling of the absorption cavity. This study finds that after the regenerative cooling, the maximum temperature of the secondary concentrator is reduced from 2400 K to 1000 K, which reaches the safe operating temperature of the secondary concentrator, while the wall temperature of the absorber is maintained above 2400 K, with very little negative effect on the heating of the propellant in the heat exchanger core. The important influence of the absorption coefficient of the RSC material on the radiation heat transfer process is analyzed. With the increase in the absorption coefficient, the RSC temperature continues to increase. The spectral band approximation model (SBAM) is used to simulate the temperature distribution characteristics of the RSC, and the calculated RSC temperature is lower.

A high-efficiency heat exchanger core of the solar thermal thruster is designed with a laminated structure; the fluid-solid coupling heat transfer method is used to simulate the laminated heat exchange channel, and the temperature of the heat exchanger core at the outlet of the laminate exceeds 2200 K. The effect of the laminated structure parameters on the heating effect of the heat exchanger core is analyzed, and the obtained main design criteria are as follows: (1) the throat area of the nozzle must be the minimum cross section in the thruster runners, which requires the sum of control runner cross sections in the laminated heat exchanger core to be larger than the throat area; (2) the optimal length of control runner is approximately 50% of the total length; (3) the smaller the cross-sectional area of the control runner is, the better the heating effect is.

As a propellant, hydrogen has a low storage density and needs for cryogenic storage, which makes hydrogen unsuitable for space applications of small spacecraft such as small satellites. After ammonia dissociation, nitrogen and hydrogen are generated, and the specific impulse can be increased to more than 400 s, which makes ammonia a great choice for a propellant. Based on the finite-rate chemistry (FRC) model, the heat transfer and flow characteristics of a solar thermal thruster using ammonia as a working fluid are investigated, with a focus on analyzing the component characteristics and variation patterns of ammonia dissociation after being heated to a high temperature and the effect of ammonia dissociation reaction on the specific impulse of the thruster. The main components of the dissociated ammonia working fluid are N_2 and H_2. The molar percentages of the dissociated components N, H, NH, NH_2, NNH and N_2H_2 in the final products are very small, and these dissociated components do not affect the performance of the thruster nozzle. Since these components are important intermediates for the formation of the final products N_2 and H_2, their role cannot be ignored. The specific impulse of the thruster after ammonia dissociation increases significantly, and the specific impulse increases with increasing ammonia dissociation; the specific impulse of ammonia can reach 410 s when the ammonia is completely dissociated. Therefore, if the thruster materials allow, the heating temperature of the heat exchanger core should be increased as much as possible to increase the ammonia dissociation, thus increasing the specific impulse of the thruster.

This book is not only a summary of our long-term engagement with STP systems research but also an inductive reflection after referring to many relevant Chinese and foreign books. Since the performance optimization/analysis of STP systems is a complex research topic, the working mechanisms of many components are still unclear. Since the performance optimization/analysis of STP systems is still under continuous development and changes, there are bound to be many limitations in this book, and any opinions and feedback are appreciated!

This book was supported by the National Natural Science Foundation of China (No. T2221002) and Natural Science Foundation of Hunan Province (No. 2024JJ6456).

Changsha, China

Minchao Huang
Jianjun Wu
Jian Li
Yuqiang Cheng

Introduction

This book takes a solar thermal thruster as the research object, establishes and expounds the mathematical model, the simulation model and experimental methods for the solar thermal thruster. The main contents include the following: it introduces the physical model and basic methods for the solar thermal thruster, carries out the researches on the radiation heat transfer of the absorption cavity and the regenerative cooling of the secondary concentrator, implements the simulation and optimization design analyses of the laminated heat exchanger core. It is also performed on the simulation analysis of the dissociation characteristics of ammonia propellant and conducted on heating experimental studies of solar thermal thruster. The above theoretical analyses or experimental studies reflect the latest research results on the performance analyses of the solar thermal thruster.

This book can be used as a textbook or reference for teachers, students and scientific personnel in aerospace, aeronautics, engineering thermophysics and power fields engaged in the performance analysis or experimental study of the solar thermal thruster.

Contents

About the Authors

Minchao Huang, Ph.D., is a professor at College of Aerospace Science and Engineering of National University of Defense Technology (NUDT) in China.

Jianjun Wu, Ph.D., is a professor at College of Aerospace Science and Engineering of National University of Defense Technology (NUDT) in China.

Jian Li, Ph.D., is a lecturer at College of Aerospace Science and Engineering of National University of Defense Technology (NUDT) in China.

Yuqiang Cheng, Ph.D., is a professor at College of Aerospace Science and Engineering of National University of Defense Technology (NUDT) in China.

Chapter 1
Preface

1.1 Research Background and Significance

The continuous development of space exploration technology requires more efficient and economical space transport systems, and there is an urgent need for propulsion systems that are more efficient than chemical propulsion. Solar thermal propulsion (STP) systems have a relatively small size and high specific impulse, thus offering performance advantages in specific space missions. An STP system is generally composed of a concentrator, heat exchanger core, nozzle, and propellant supply system. The concentrator typically uses parabolic mirrors to concentrate sunlight to heat the heat exchanger core at a focal point. The propellant is heated when it flows through the heat exchanger core and is finally expanded and accelerated by a Laval nozzle to generate thrust. In this propulsion mode, the propulsion system can generate a Newton-level thrust by using an expandable whirling-membrane concentrator with an area of approximately 10 m^2 to collect solar radiation. Because hydrogen has a small molar mass and can generate a high exhaust velocity, it is an ideal propellant for solar thermal thrusters. The theoretical specific impulse of an STP using hydrogen as a propellant can reach 800 s [1]. However, the low molar mass of hydrogen also reduces the thrust that the engine can provide, limiting the STP system to low-thrust space missions. The main applications of STP systems are orbit transfer and interplanetary exploration, which require a thrust level between 0.1 and 40 N. Figure 1.1 compares the thrust ranges and specific impulse ranges of several common propulsion methods. Chemical propulsion has a large thrust and a low specific impulse, and electric propulsion has a high specific impulse but a very small thrust. Within the size range of currently achievable spacecraft platforms, the specific impulse and thrust characteristics of STP systems are very attractive, and their high specific impulse and moderate thrust fill the gap between chemical and electric propulsion systems; as such, they are expected to be used to increase the payload ratio of orbital transfer vehicles (OTVs) or orbital maneuvering vehicles (OMVs).

M. Huang et al., *Solar Thermal Thruster*, https://doi.org/10.1007/978-981-97-7490-6_1

Fig. 1.1 Comparison of thrust ranges and specific impulse ranges of several common propulsion methods

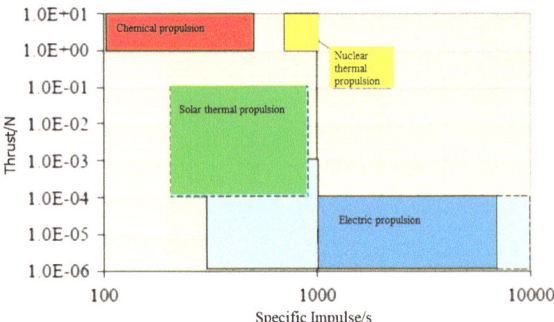

Traditional chemical propulsion technology uses chemical energy to send a vehicle into a predetermined space orbit or achieve in-orbit maneuvering of a spacecraft, primarily by using liquid and solid propellants. Chemical propulsion has a development history of nearly 100 years. At present, the theoretical systems and applied technologies of chemical propulsion are basically mature, and supporting facilities, such as launching bases and ground measurement and control systems, are sound. The most prominent feature of chemical propulsion is that it can provide a large thrust, and chemical propulsion has always been the most used propulsion technology in the aerospace field and will be the most important space propulsion technology in the foreseeable future. Although traditional chemical rocket propulsion technology can meet the needs of current launching tasks in terms of functionality, safety, and reliability, with the dramatic increase in commercial launches and the expansion of space exploration missions, there are new requirements for launch costs, launch cycles and payload capabilities. STP technology can increase the payload into orbit, reduce the mass of power systems for orbit-holding and maneuvering, shorten the transfer time from near-Earth orbit to geosynchronous orbit, and greatly expand human interplanetary exploration capabilities. STP is a cutting-edge space propulsion technology, and its functions include primary propulsion, response and adjustment, position keeping, precise pointing, and orbital maneuvering. The primary engine used in a space mission provides the main thrust for orbit transfer, interplanetary orbit, and landing and ascent on exoplanets. The response and adjustment and orbital maneuvering systems provide thrust for orbit-holding, azimuth control, position keeping and spacecraft attitude control.

Launching from the Earth's surface requires a thrust-to-weight ratio greater than 1. Currently, chemical propulsion is the only propulsion technology that can generate thrust that overcomes the Earth's gravity. In space, a propulsion system with higher efficiency can be used to reduce the total mass of propellant needed for the mission. The limitation of chemical propulsion lies in its relatively low specific impulse. Many advanced propulsion technologies have higher specific impulses, such as electric propulsion, which is widely used in the position keeping of commercial communication satellites and the primary propulsion of some scientific missions. However, the thrust from electric propulsion is very low, and therefore, it takes a long time to provide the total impulse needed for the mission. STP can provide a Newton-level thrust, so the time to complete tasks such as orbit transfer can be significantly shortened compared to electric propulsion.

At present, the United States (US), Japan, Russia, and the United Kingdom (UK) are all conducting research on STP-related technologies, and some have entered the test and improvement stage from the program demonstration and prototype development stage [2–4]. The National Aeronautics and Space Administration (NASA)'s Advanced Propulsion Technology and Development Program also lists STP technology as a near-term priority and achievable project. To date, the heating methods of STP are generally indirect; that is, concentrated sunlight is used to heat a solid, and then the red-hot solid is used to heat fluid to a high temperature above 2000 K. For the STP technology, there are three key components: large light concentrators, high-temperature-resistant secondary concentrators, and high-efficiency heat exchanger cores. The main technical difficulty of the heat exchanger core is the high-temperature structure, including preventing it from reacting with the working fluid at high temperature, maintaining a stable configuration for high-speed and efficient heat exchange with the working fluid, maintaining the material strength at high temperature, and reducing the external radiation heat transfer. How to further enhance the heat exchange structure of the STP system and improve the integrated design of the thruster are issues that have attracted extensive attention from researchers around the world [5].

In China, the Beijing University of Aeronautics and Astronautics, Harbin Institute of Technology, Northwestern Polytechnical University and National University of Defense Technology have also conducted basic theoretical research on STP systems, mainly focusing on system conceptual designs and flow simulations, and the relevant research needs to be further improved [8].

1.2 Research Overview

1.2.1 Structural Design of Solar Thermal Thruster

In the US, the Phillips Laboratory operated by the US Air Force Materiel Command, NASA Marshall Space Flight Center, NASA Lewis Research Center, and the University of Alabama are conducting research on STP technology [14–18]. The key research projects include the Integrated Solar Upper Stage (ISUS) and Solar Orbit Transfer Vehicle (SOTV) [19–21]. Figure 1.2 shows a conceptual diagram of the application of a STP system to a space vehicle.

Rockwell (USA) developed an early initial prototype of a rhenium thrust chamber, while the Pratt & Whitney Rocketdyne (USA) developed a thrust chamber using a porous material composed of hafnium carbide (HfC) [22]. In 1996, a scheme demonstration and test prototype study of an ISUS propulsion system was started using a tungsten or tungsten alloy absorber/thrust chamber and hydrogen as the

Fig. 1.2 Conceptual
diagram of the application of
an STP system to a space
vehicle

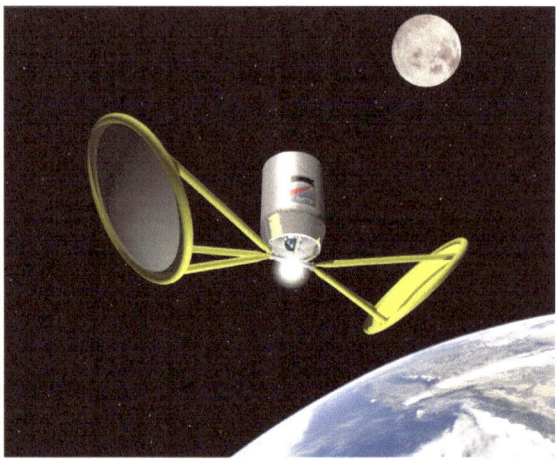

working fluid [23]. The Marshall Space Flight Center (MSFC) (USA) has been
developing solar thermal thrusters. Its most typical solar thermal thruster is designed
for a thrust of 2 lb (approximately 8.9 N), a specific impulse of 860 s, and a thrust
chamber temperature of 2533 K. The absorber wall is cylindrical with a thickness of
0.2 in and a length of 16.3 in, the bottom is hemispherical with an opening diameter
of 2.652 in and a bottom diameter of 2.588 in, and the bottom is narrower than
the opening to prevent the absorber from deviating from the central axis. The outer
diameter of the housing is 3.213 in and the bottom is 3.17 in, and the wall thickness
is 0.2 in, as shown in Fig. 1.3.

In 1997, the NASA Lewis Research Center conducted ground experiments on
the ISUS system, with a focus on the absorber/thrust chamber (Receiver/Absorber/
Converter, RAC) [24–28]. The RAC is a graphite cavity, and to avoid a reaction with
hydrogen, the inner and outer surfaces are treated by a chemical vapor deposition of

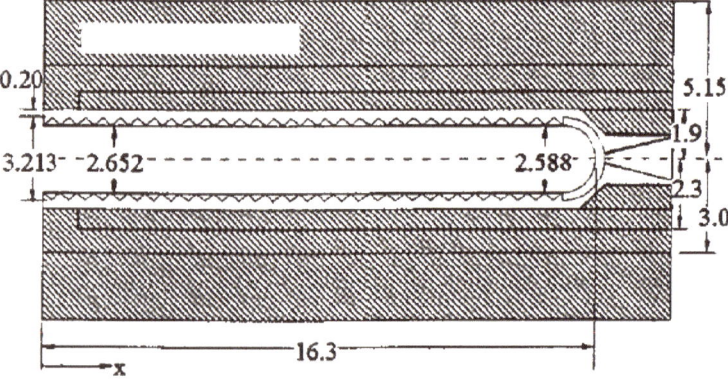

Fig. 1.3 MSFC solar thermal thruster structural diagram

rhenium. The working fluid H_2 enters the RAC after being preheated in the preheater. In the experiment, the measured maximum wall temperature of the RAC was 2200 K, the maximum temperature of the working gas was 2022 K, the flow rate of the working gas was 1.7 g/s, and the calculated specific impulse reached 742 s. The parameters of the ground demonstration test of the solar thermal thruster are shown in Table 1.1. This series of tests successfully verified the feasibility of STP technology at the system level. The solar thermal thruster its working medium in hot runner USES the traditional simple structure of the spiral flow channel. The working fluid heat exchange runner uses a traditional spiral runner with a simple structure. Another heat exchange structure has 195 fine flow channels with a diameter of 3 mm and a length of 16 cm drilled on a graphite cavity along the axial direction. After the test, the thrust chamber and multilayer insulation (MLI) protective layers were inspected, and no abnormalities were found. The MLI technology and spiral runner design of the thrust chamber can be used for reference and improved in follow-up studies. The MLI package is composed of 80–100 layers of tungsten and molybdenum foils, which can effectively reduce the radiation of the thruster to the external environment. In the experiment, the energy loss of the thruster was measured to be approximately 4800 W (input power 10 kW), with radiation loss accounting for 80% and convective loss accounting for 20%. A spiral runner is a measure often used in engineering to enhance heat transfer. During the forward movement of the working fluid in the spiral runner, the direction is continuously changed, which could cause secondary circulation in the cross-section and enhance heat transfer. The comparison of the temperature difference between the heat exchanger core outlet airflow and the inlet airflow and the temperature difference between the heat exchanger core body and the inlet airflow confirmed that the heat exchange efficiency was approximately 90%.

In 2004, a physical science company received funding from the U.S. Air Force Research Laboratory and cooperated with Rocketdyne and Boeing to develop an innovative STP system for small aircraft [29–30]. In this system, the solar radiation is concentrated by the concentrator, and the high-intensity solar radiation is transmitted to the thermal absorber through low-loss fiber optic light guides to generate effective and high-performance thrust. A graphite absorber is used in this system, and the outside is insulated by molybdenum. The working fluid flows into the cavity through the small holes of the graphite cavity and then flows out of the heating cavity through molybdenum tubes, where it is fully heated through tortuous flow paths. This design

Table 1.1 Parameters for the ground demonstration test of the STP system	Ground demonstration test parameters	Test value
	Highest temperature of absorber/K	> 2200
	Highest temperature of working fluid/K	> 2012
	Corresponding specific impulse/s	742
	Heat exchanger efficiency/%	– 90
	Number of ignitions	122
	Total time/h	320

concept is used for reference in the design of the working fluid flow channel in the thrust chamber, which makes the working fluid flow in a circuitous manner inside the high-temperature wall.

Molybdenum and tungsten and their alloys have high melting points and are ideal materials for solar thermal thrusters. However, they can become brittle during processing, welding and high-temperature recrystallization, so the manufacturing of thruster materials using molybdenum or tungsten is difficult. The Japan Aerospace Exploration Agency, in cooperation with the Japan Science and Technology Agency and the National Institute of Materials Science (NIMS), doped molybdenum or tungsten billets with a small amount of CaO and MgO and then hot-rolled them into single-crystal molybdenum or tungsten plates to improve the ductility of molybdenum and tungsten and their alloys, which solved the brittleness problem of molybdenum, tungsten and their alloys during processing, welding and high-temperature recrystallization [31–35]. Large and medium-sized STP devices were successfully developed using single-crystal molybdenum, small STP devices with back-to-back arrangements of dual absorbers/thrust chambers were fabricated using single-crystal tungsten, and preliminary performance experiments of STP devices were conducted using N_2 or He as the working fluid. After the sunlight is concentrated by the concentrator, it directly heats the thruster. Layered heat insulation technology is used on the outside of the thruster. A concentrator with an opening diameter of 1.6 m is used, the heating temperature of the working fluid reaches 2300 K, and the specific impulse with H_2 as the reference reaches 800 s. as shown in Fig. 1.4. At present, molybdenum and tungsten production in China can meet the working requirements of the thruster through specific processing technology. This book preliminarily considers the use of molybdenum or tungsten alloys for the thruster.

Li et al. [36] proposed an innovative design of Fresnel concentrator for Space Solar Power Satellite (SSPS) based on fiber bundles, as shown in Fig. 1.5. It includes planar Fresnel lens arrays for solar concentration, fiber optic bundles for transmitting concentrated sunlight to photovoltaic panels, and highly modular interlayer modules

Fig. 1.4 Typical single-crystal coaxial NAL STP thrusters

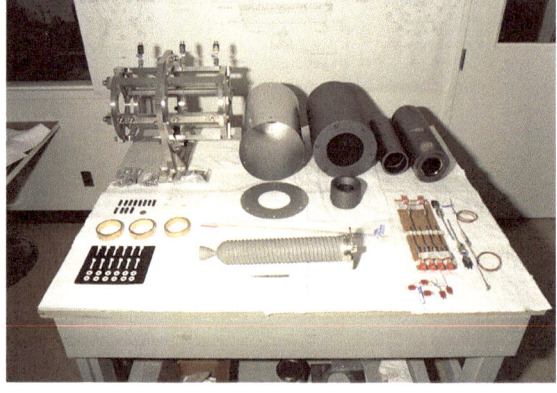

Fig. 1.5 Fresnel concentrator based on fiber bundle SSPS design

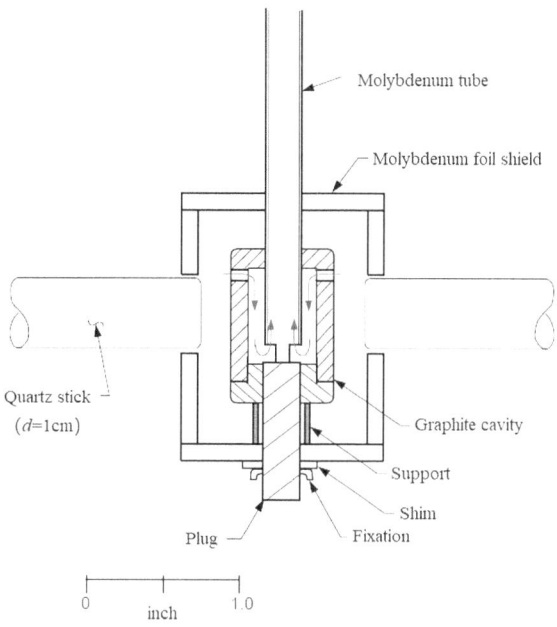

Molybdenum tube

Molybdenum foil shield

Quartz stick (*d*=1cm)

Graphite cavity

Support

Shim

Plug

Fixation

0   inch   1.0

for power generation/transmission. In addition, designed with secondary homogenizer Fresnel concentrator, to obtain high concentrated on solar equipment than the irradiance uniformity and good. The thermal analysis was carried out on the mezzanine module, in order to verify whether the temperature fluctuations remain within the design requirements, the results prove the feasibility of SSPS design. However, for very large scale of Fresnel lens array attitude control and maintenance, still need to complex configuration, such as satellite platforms more forms and more advanced technology, as well as some effective deployment in space, the assembly or manufacturing methods.

Zhou et al. [37] proposed a compact solar concentrator with integrated prism and semi-parabolic trough mirror for efficient collection of solar energy in a limited space, as shown in Fig. 1.6. The rotating prism is used to track sunlight, and the incoming light is perpendicular to the aperture of the semi-parabolic groove mirror. The radiant energy from the prism is then reflected and concentrated by the semi-parabolic trough mirror, and finally absorbed by the receiving tube. Based on a strict theoretical model, the tracking strategy is proposed and verified. The results show that the tracking strategy can accurately track the sun under the condition of non-parallel multispectral solar radiation. The geometric concentration ratio, geometric parameters and row arrangement of the proposed solar concentrator are optimized. The results show that the estimated annual solar thermal efficiency of the concentrator is as high as 41.1%, which is 6.7% and 17.6% higher than that of the parabolic trough concentrator and the flat plate concentrator, respectively. Therefore, the concentrator has a better application prospect.

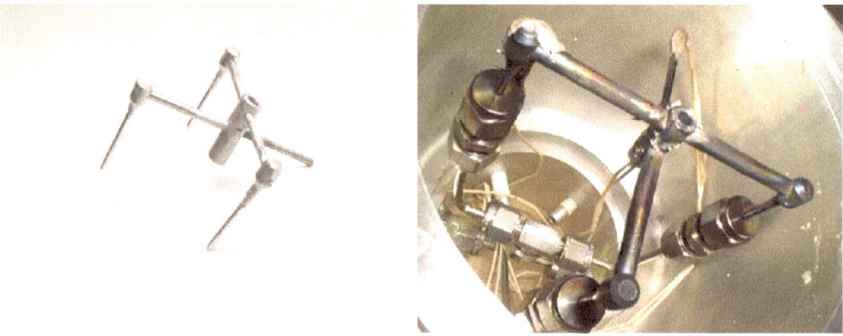

Fig. 1.6 Parabolic trough solar condenser

Fan et al. [38] Aiming at the problems existing in the original SSPS concept, a spherical secondary condenser design is proposed, as shown in Fig. 1.7. Bessel curve is used to describe the geometry of the bus of the secondary condenser, and Monte Carlo ray tracing method is used to evaluate the optical properties and energy density distribution of the array. Then, a mathematical optimization model is established and particle swarm optimization algorithm is used to obtain better secondary condenser configuration, solar energy collection performance and energy density distribution. The results show that when the order of Bessel curve is 5, the effective solar energy collection and solar energy collection can reach 81.14% and 87.06% respectively, and the energy density distribution can be further optimized, and the relative unhomogeneity index of 0.4 can be achieved.

The U.S. Air Force Phillips Laboratory has completed testing of spiral-rhenium tubular heat exchangers and thrusters [39]. The solar thermal propulsion unit is capable of withstanding 20 kW of thermal power at a design temperature of 2778 K

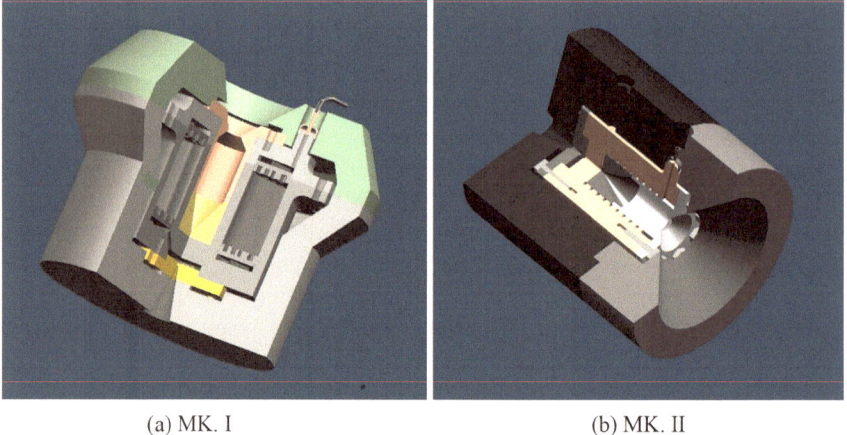

(a) MK. I   (b) MK. II

Fig. 1.7 Spherical solar concentrator

and producing a specific pulse of 810 s. The test device successfully operated for 65 h, reached the highest gas temperature of 1810 K, and the specific impulse produced could reach 650 s.

The surface area ratio of the absorber to the inlet should be large enough to make the absorption cavity approach a blackbody. As planned, the surface area ratio of the absorption cavity to the inlet of the US ISUS should reach 50:1 [40]. At present, there are two main modes of STP: (1) particle bed pulsed thrust (PBPT), and (2) jacketed continuous thrust (JCT). The ideal temperature distributions of the two modes are different. The JCT mode requires the maximum energy flow to be distributed at the outlet so that the working fluid can obtain the highest temperature there. The improvement in PBPT performance lies in the rapid and uniform heating of the entire absorber. Therefore, the design of the concentrator should be based on the different requirements of the propulsion mode to make the distribution of the light spots more reasonable and more conducive to improving the performance of the thruster. The shape and structure of the absorber also require different adjustments [41]. The surface structure of a material directly affects the absorption, reflectance, and emissivity. The reflectance of the absorber material is a key factor affecting the performance of the absorption cavity. Therefore, the inner wall of the absorber needs to be polished or roughened to meet the needs of different propulsion modes (polishing corresponds to the PBPT mode, and roughening corresponds to the JCT mode). Surface roughening can increase the energy absorption rate of the inner wall of the absorption cavity, which is mainly achieved by surface knurling treatment and the addition of a carbon black chip coating [42]. At present, improving the processability of materials and components is a key focus of absorber/thrust chamber research.

In 1996, Denise Stark et al. of the University of Alabama [43] performed experimental evaluations on absorber and concentrator surfaces. For the PBPT and JCT modes, many tungsten absorber samples were processed, the inner wall surfaces were polished and roughened to different degrees, and the temperature distributions of various absorbers were tested using high-power concentrated light. The experimental results showed that after surface treatment, the heating effect and performance of the corresponding propulsion mode can be greatly improved. For the current research plan, the JCT mode is mainly considered, the absorber structure is designed to be conducive to sunlight absorption, and the surface is roughened.

Henshall from the University of Surrey established a ray trace model inside an absorber based on fiber transmission and analyzed the light intensity distribution and temperature gradient at the absorber wall [44]. In the simulation, it was assumed that the interior of the absorber was totally reflective (100%), and monospectral light was used. For the absorber configuration, the simulation results showed that the light intensity at 5 mm along the wall is the maximum, and the temperature is also the highest, as shown in Fig. 1.8. This analysis has important guiding significance for the design of an optimal absorber structure.

To achieve a temperature above 2300 K for the propellant, the concentration ratio of the concentrator needs to reach approximately 10,000:1. However, a single primary concentrator can hardly meet this requirement, so the concentrator system generally needs to be equipped with secondary concentrators. At present, there are

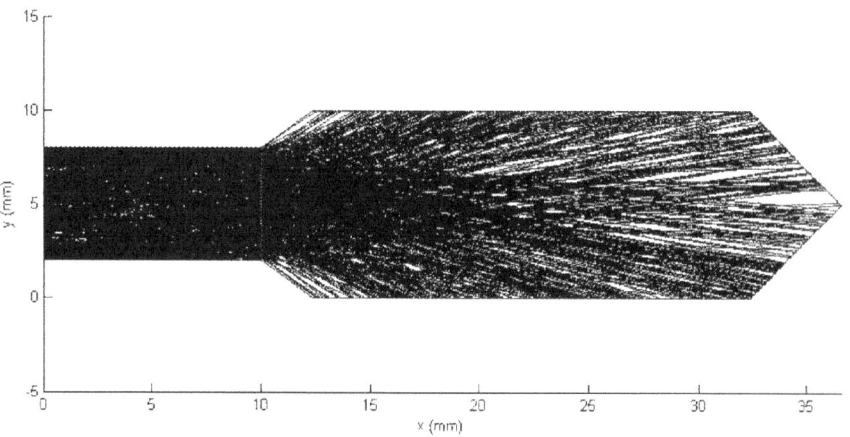

Fig. 1.8 Ray trace model of the interior of the absorber

two main design schemes for secondary concentrators, namely, compound parabolic concentrators (CPCs) and reflective secondary concentrators (RSCs) [45–47]. Due to the considerable reflection and absorption losses of secondary CPCs, their output efficiency is only approximately 65%. Compared with secondary CPCs, the greatest advantage of RSCs is their high transmission efficiency. To reduce the requirement of the primary concentrator on solar tracking accuracy, compensate for the resulting loss of the concentration ratio, and improve the efficiency of the thrust chamber in absorbing solar radiation as much as possible, an RSC is used as the secondary concentrator in the STP system.

The RSC is prone to cracking at high temperature. The NASA John H. Glenn Research Center conducted high-temperature tests on two sapphire RSCs in 2009, and both sapphire RSCs separately cracked at 1300 and 649 °C, as shown in Fig. 1.9 [48]. Based on the measurement of the stress on the cross-section, the radial tensile stress on the lens surface reached 44–65 MPa, and the analysis revealed that the damage to the lens surface during processing and operation caused cracking. The second concentrator was broken into two parts and discolored, i.e., the energy extractor turned silver–gray, and the lens turned brown. When an RSC is used in a high-temperature environment, it is prone to cracking. The cracking is random, which requires strict processing requirements, and it is very difficult to manufacture a reliable and durable RSC. Integrated active cooling measures can be used to improve the service life of RSCs.

Many innovative studies on STP technology have also been conducted in China. Dai et al. conducted a preliminary simulation study on a photothermal conversion mechanism, used the Monte Carlo method to trace rays, and analyzed the effect of a quartz window on optical radiation absorption [49–53]. Only when the heat conversion temperature is higher than the critical temperature under given concentration conditions can the quartz window improve the heat conversion efficiency. The higher

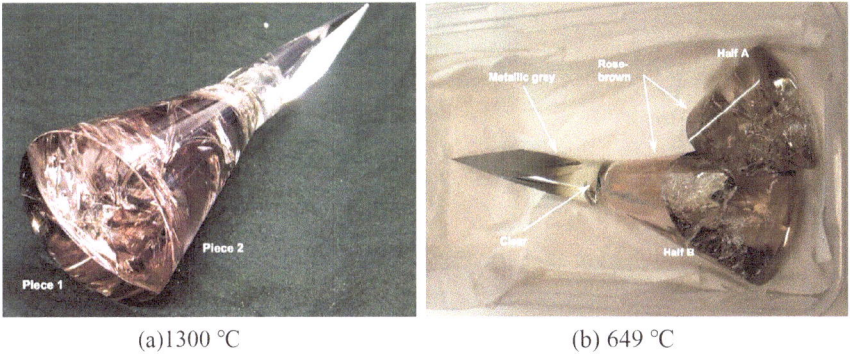

(a)1300 °C   (b) 649 °C

Fig. 1.9 Photos of ruptured RSCs under high-temperature tests at the NASA John H. Glenn Research Center

the concentration ratio of the incident solar energy, the higher is the critical temperature. The temperature distribution of the heat absorption cavity has a significant impact on the heat conversion efficiency. The farther the temperature distribution peak moves from the solar incident window, the higher the efficiency. Dai et al. provided great reference value for absorber design and analysis.

1.2.2 High-Efficiency Heat Exchange Runners for Solar Thermal Thrusters

An efficient heat exchange runner is the key to achieving a high specific impulse for an STP system. The specific impulse of the propulsion system can be increased by increasing the temperature of the propellant entering the nozzle. Because of its simple structure and easy processing, the spiral runner is the most used structure in foreign studies [54]. An increase in the length of the thrust chamber leads to a large system volume, a complicated structure, and an increased mass, which reduces the ability of the thruster to complete its tasks.

Ma et al. from Zhejiang University [57] studied the flow structure and heat transfer characteristics in a rotating curved circular tube in depth and analyzed the effects of different cross-sectional shapes and geometric parameters of the tube at the inlet section and the fully developed section on the axial velocity distribution, cross-sectional secondary flow structure, turbulent kinetic energy, friction ratio of the curved tube, temperature distribution and Nusselt number. Their results provide guiding significance on the design of a spiral runner as the heat transfer structure for a solar thermal thruster. However, spiral runners still suffer from low heat transfer efficiency.

With reference to the transpiration cooling of laminated structures, this book uses a laminated runner with high-efficiency heat transfer to increase the heat transfer area between the working fluid and the thrust chamber wall through flow shunting so that the working fluid can be fully heated in the thrust chamber. At the same time, the working fluid flows from the low-temperature outer wall to the high-temperature inner wall along the radial direction, which reduces the temperature of the outer wall, reduces the heat loss of the thruster to the external environment, protects the wall surface material, and improves the energy use efficiency of the system.

At present, laminate technology is mainly used for transpiration cooling in thrust chambers of liquid propellant rocket engines. Laminated transpiration cooling can limit the heat-affected zone to the dispersed flow zone, and due to the high friction of the control runner and the low friction of the dispersed flow zone, the effect of local overheating of the heated wall on the control runner is small. Even when local overheating occurs, the flow rate of coolant is basically constant, and the high-temperature region due to the local overheating can be cooled to a normal level under the action of stable transpiration cooling. Therefore, the possible local overheating of general porous materials by transpiration cooling can be overcome, thus achieving the reuse of heated parts. Due to the advantage of the laminated structure, the efficient convective heat transfer between the working fluid and the high-temperature laminates becomes safer and more reliable.

The US Aerojet Corporation has logged many achievements in laminate technology. In the mid-1990s, an attempt was made to develop a novel dual-fuel double compression–expansion engine. The head of the thrust chamber used a laminated mutual impinging jet injector. After the control runners and dispersed flow runners for the coolant were etched out of a metal sheet, diffusion welding technology was applied to form a nozzle with the laminated structure, which achieved a design scheme of transpiration cooling of the thrust chamber wall with H_2 as the coolant. A ground thermal test confirmed that the use of tiny gaps in the metal laminates for transpiration cooling can precisely control the transpiration cooling flow rate of the propellant in the tiny gaps of each laminate, and a very small amount of H_2 as coolant can achieve effective cooling of chamber wall, which pushed the research and application of laminated transpiration cooling technology to a new stage [60].

In China, research on the transpiration cooling technology using various types of laminates began in the 1990s, with the main research focused on the thermal protection of high-temperature aerospace devices.

Yu et al. from Northwestern Polytechnical University [62–64] conducted experimental studies on the flow resistance characteristics of a scaled-up laminate model in suction type and blowout type wind tunnels based on the similarity principle, analyzed the influences of internal structural parameters of the laminates, including open area, spoiler column shape, channel height, and spoiler column arrangement, on the flow resistance characteristics, established an engineering calculation model for the flow resistance characteristics of the laminated structure, studied the internal heat transfer characteristics of some structures, and obtained the internal heat transfer characteristics of some laminated structures. Yu et al. also used numerical simulations to investigate in detail the complex flow and heat transfer patterns within the laminates. Quan et al. [65] used experiments and numerical simulations to systematically study the laminated cooling technology of turbine blades, designed and fabricated laminated test pieces with different open areas and internal structures, tested the flow resistance characteristics and cooling effectiveness in a modified return wind tunnel, and investigated the effects of the open area and shape of the spoiler post on the flow resistance characteristics. Quan et al. designed and manufactured inlaid laminated cooling blades with different turbulences, conducted experiments on the flow resistance characteristics and cooling characteristics in a small cascade wind tunnel for heat transfer, and investigated the internal flow and heat transfer patterns of the laminated experimental parts under experimental conditions and real working conditions of the blades by using the fluid–solid coupling heat transfer method.

Niu et al. [68] from Shanghai Jiaotong University studied the regenerative cooling technology of liquid propellant rocket engines using laminates, established a mathematical model of the 2D fluid–solid coupling heat transfer process in a thrust chamber with regenerative cooling using laminates, investigated the 3D effect of the turbulent flow heat transfer in a liquid propellant rocket with regenerative cooling using laminates, proposed a 3D flow heat transfer calculation method in the regenerative cooling channel of a liquid propellant rocket engine, and used this method to numerically simulate the heat transfer process of regenerative cooling in a scaled-down thrust chamber. For the first time in China, Niu et al. experimentally investigated the flow and convective heat transfer characteristics of a regeneratively cooled channel using large aspect ratio laminates, providing the necessary theoretical methods and preliminary experimental basis for the design of a large aspect ratio regeneratively cooled channel of the thrust chamber using laminates. Yang et al. [69] explored the basic theory and practical application of laminated transpiration cooling from the two perspectives of experimental research and theoretical analysis, conducted experiments on the flow resistance characteristics of microchannels with curved rectangular cross-sections adjusted by an adjustment plate of the laminated transpiration coolant, studied the flow and heat transfer characteristics of coolant dispersed runners with different geometric sizes, and established a numerical simulation of the flow resistance characteristics of the adjustment channel. Yang et al. used the fluid–solid coupling heat transfer method to analyze the heat transfer characteristics of laminates in a thrust chamber with transpiration cooling, proposed the concept of a thrust chamber with regenerative-transpiration dual-mode cooling, and offered a scheme design, analysis and demonstration.

Based on an analysis of research methods on the temperature fields of porous materials and laminated structures with transpiration cooling, Liu et al. from the National University of Defense Technology [71] obtained a series of algorithms for the steady temperature field of a thrust chamber with laminated structures by reasonably processing the boundary conditions, proposed a computational model and calculation method for the unsteady 2D axisymmetric temperature field of a whole chamber wall, and obtained rapid estimation, analytical calculation and numerical calculation methods for a chamber wall temperature field with laminated transpiration cooling. In addition, to apply the stability theory to thin plates, Liu et al. proposed that the laminated structure of a transpiration-cooled thrust chamber could suffer thermal damage under certain working conditions. In 2008, based on microscale theory, Zhang [73] numerically analyzed the heat transfer characteristics of dispersed runners, the evaporation and combustion rates of beaded transpiration of staggered pores on the laminates, and the flow and combustion characteristics of a transpiration medium in a staggered layered transpiration pore structure, performing studies on thermomechanical coupling characteristics and test validations of typical laminated structures with transpiration cooling to provide a theoretical basis and reference for guiding the design of related mixers and combustion devices.

Control and dispersion laminates are usually very thin, and there are microscale effects on flow and heat transfer in the channel. When the coolant inlet pressure is constant, the coolant flow rate can be regulated by adjusting the actual flow distance of the coolant in the primary and secondary control runners. The heat transfer of high-temperature gas to the transpiration coolant is mainly performed in the dispersion runner. Therefore, investigating the microscale effects of flow and heat transfer in laminated channels and studying the flow resistance characteristics of control runners and the heat transfer characteristics of coolant in dispersion runners are important for the safety of a thrust chamber. In microscale systems, the phenomena exhibited by fluid flow and heat transfer are quite different from macroscopic large-scale flows, and the interactions between fluids and surfaces are also considerably different, which have garnered increasing attention from researchers in China and abroad. When the characteristic size of a flow field is at the micron or submicron level, the velocity slip and temperature jump, as well as thermal creep, electrokinetic effects, viscous heating, anomalous diffusion or even quantum theoretical and chemical effects, may dominate the fluid, resulting in anomalous microscale flow phenomena. In general, the main factors affecting microscale flow and heat transfer are different for gases and liquids. For gases, there are four important influencing factors for the anomalous phenomena of flow and heat transfer: rarefaction effects, compressible effects, viscous heating, and thermal creep. When studying microscale gas flow characteristics, the rarefaction effect of gas is the main influencing factor, while the flow of microscale liquid is mainly affected by the surface force and intermolecular force. Therefore, the influence of microscale effects needs to be considered in these microscale systems. Some factors that are negligible at the macroscale become main influencing factors in a microscale system. For example, the ratio of surface area to volume is a very important factor in the study of heat transfer; the surface tension between a liquid and a solid wall has a direct impact on the flow pattern and rate of liquid in microchannels.

In general, the structure of microchannels is relatively simple, and the internal flow is mainly laminar flow and can appear as turbulent flow at high Reynolds numbers. At the beginning of the twentieth century, Knudsen, Gaede et al. conducted experiments on gas flow in microchannels. However, because of the low precision of experimental equipment and the limitations of experimental methods at that time, the study of microscale channels was more of a qualitative analysis. With continuous improvements in microelectromechanical system (MEMS) technology since the late 1980s, researchers have conducted more detailed studies on the flow and heat transfer characteristics in microchannels [74]. In the early 1990s, Pfahler et al. and Harley et al. conducted experiments on gas flow in microchannels with Reynolds numbers in the range $0.5 \leq Re \leq 20$ and analyzed the effect of gas rarefaction on the outlet pressure drop and friction coefficient [75]. In 1993, Arkilic and Breuer [77] conducted experiments on the flow of argon gas in a microchannel with L = 7.5 mm, W = 52.25 μm, and H = 1.33 μm and obtained a set of high-precision data.

In 1995, Jiang from Tsinghua University [79] performed an experimental study of liquid in microscale straight channels with a diameter of D = 8–42 μm; the experimental results matched the theoretical calculations, and the flow conformed to the macroscopic flow pattern. In 2001, Qin et al. from the University of Science and Technology of China [80] conducted experiments on the flow of nitrogen and helium in microcircular tubes with a diameters of D = 17.6–17.9 μm and lengths of L = 10–70 mm, and the compressibility of gases at a low Mach number was investigated. In addition, academician Zengyuan Guo of Tsinghua University [81] has been committed to the study of microscale flow and heat transfer problems. In 1997, Xiaobo Wu and Zengyuan Guo used numerical calculations to study the flow and heat transfer characteristics in microtubes and proposed that the effect of fluid compressibility on the velocity profile should be considered [82]. In 1999, Du et al. [83] used numerical calculation to investigate the effects of pressure work and viscous dissipation on the adiabatic flow characteristics of compressible fluids in microtubes.

Boyd, Fan and Shen from the Institute of Mechanics of the Chinese Academy of Sciences conducted an exploratory study on the gas flow in a microchannel using the improved direct simulation Monte Carlo–information preservation (DSMC-IP) method [84]. In 2004, Boyd et al. numerically simulated the flow of rarefied gas in a microchannel using a continuous medium coupled with DSMC-IP and found that the calculation results of the coupling method were significantly better than those obtained by the continuous medium alone or the DSMC method alone.

In 2007, Qi from the Chinese Academy of Sciences [87] used the DSMC method to simulate the flow and the heat transfer of a single-component gas and a two-component gas mixture at the microscale, discussed the physical mechanism from the perspective of molecular motion, and analyzed gas flow at the microscale and rarefied gas flow at the conventional scale. To ensure the validity of the DSMC method in microscale studies, the flow similarity condition was discussed. Numerical simulations were performed for the flow in the microchannels and the flow in the square cavity, and the basic flow problems were discussed and analyzed. On this basis, the flow of the two-component gas mixture was studied, and the results showed that the entry velocity of the gas had a significant impact on the mixing distance of the two gases and had a great impact on the proportion of a single gas after being completely mixed.

In 2010, based on the concept of multirelaxation lattice Boltzmann models, Zheng of the Huazhong University of Science and Technology [88] improved the lattice Boltzmann method (LBM) in heat and mass transfer, studied the two key issues of the LBM for microscale heat and mass transfer, and conducted useful explorations of the frontiers of microscale heat and mass transfer. Zheng proposed 2 boundary processing formats, i.e., the boundary conditions for the combined equilibrium state and specular reflection and combined equilibrium state and rebound, analyzed the LBM boundary processing format for the flow and heat and mass transfer in a 2D or 3D microchannel, discovered the dispersion effect, and provided the corresponding correct processing format. The basic model proposed based on theory successfully captured the anomalous thermal creep phenomenon at the microscale, and it was found that the uncoupled thermal model could not capture the thermal creep phenomenon.

In summary, the relevant technologies in other countries are mature, while STP research in China has mainly focused on system design and flow field simulations, with few experimental studies. There is a lack of in-depth studies on the key light-to-thermal conversion configurations and high-efficiency heat transfer configurations. To realize the application of this technology in ISUS systems, a large amount of mechanistic research and specific innovative design work are still needed.

1.2.3 Solar Thermal Thruster Propellants

The most ideal propellant for STP systems is hydrogen because it has a molar mass of 2 g/mol, and the obtained specific impulse is significantly higher than that of other propellants under the same heating temperature. In early research on STP technology, hydrogen was mostly used as the working fluid, and for the US ISUS system, a comprehensive experimental and simulation study was conducted on hydrogen [89].

However, a large tank is needed for hydrogen storage (liquid storage density of 71 kg/m^3), and storage is difficult. The ISUS system requires a complex liquid hydrogen storage and supply system. An MLI design is used for the storage tank to prevent the pressure from deviating from the rated value, and a zero-gravity thermodynamic vent system is used to integrate the liquid acquisition instrument to achieve the subcooling effect. Therefore, a pressurization system is not needed. Liquid hydrogen is completely converted to hydrogen gas through heat exchange inside the system, so there is no two-phase flow. The system includes a liquid hydrogen regulator and preheater; the designed mass flow rate is 1.667 g/s, and the inlet pressure is 0.2 MPa. This supply system is of high mass, accounting for a large proportion of the ISUS system's mass, which is the main factor restricting the application of hydrogen as propellant [91].

Ammonia has a moderate molar mass of 17 g/mol and a storage density of 600 kg/m^3, and the technologies needed for liquid ammonia storage are simple and low in cost. Since the current liquid hydrogen storage technologies are immature, the use of ammonia as a propellant is an ideal choice. At the working temperature of the STP system, the mixture after ammonia dissociation is composed of atoms and molecules, and the existence of ions can be completely ignored due to the low temperature (less than 3000 K). For vibrational excitation, only nitrogen molecules and hydrogen molecules are considered, and they have stable vibrational excitation levels. Using the multitemperature model is an effective method, and the calculation volume is acceptable. For ionized flow in thermochemical nonequilibrium, a complete theoretical model is developed using the dual-temperature or triple-temperature model. A weakly ionized flow is often used in the simulation of nonequilibrium, ionized, hypersonic flow, and most work has studied the dissociated, ionized air flow through a dual-temperature physical model. Using a multitemperature model for numerical simulation of the dissociated flow of ammonia working fluid in a STP system is more accurate, and in-depth simulations and experimental studies on the reaction kinetics of the ammonia dissociation reaction have been conducted in other countries and provided rich dissociation reaction models.

In 1990, Davidson et al. conducted a series of high-temperature dissociation experiments of ammonia with a temperature range of 2200–3200 K and the pressure range of 0.8–1.1 atm [96]. Under this condition, ammonia was completely consumed within a residence time of 1 ms. Using the narrow linewidth laser absorption mechanism, the variation patterns of NH and NH_2, the high-temperature dissociation products of ammonia in the shock wave, were measured. A detailed mechanistic model of ammonia dissociation at high temperature was established, including 21 radical reactions, and the rate constants of the main reactions were determined. These studies are of great guiding significance for the analysis of intermediate products (NH, NH_2, N_2H_2, etc.) in ammonia dissociation.

In 2000, Konnov A A et al. [97] investigated a kinetic model of high-temperature ammonia dissociation. The key chemical reactions that determine the modeling quality were identified through sensitivity analysis, and the selection of reaction rate constants was investigated. Results showed that only a great reduction in the rate constant for the reaction $NH_3 + NH_2 \leftrightarrow N_2H_3 + H_2$ could improve the agreement of the model results with the experimental data, and in the temperature range of 2200–2800 K, the optimal rate constant for such reaction should be taken as $k = 1.0 \times 10^{11} T^{0.5} e^{-21600/(RT)}$. Results showed that if the reaction $NH_3 + NH_2 \leftrightarrow N_2H_3 + H_2$ was used and the reactions with N_2H_3 and N_2H_4 were considered in the reaction mechanism, the calculation of rise time and peak values of the NH and NH_2 components would be greatly affected. Meanwhile, since N_2H_3 and N_2H_4 account for a very small proportion of the ammonia dissociation products, reactions involving N_2H_3 and N_2H_4 were excluded.

In 1996, A. Chambers et al. studied the decomposition characteristics of ammonia in the coal gas atmosphere of a gasifier [98] and compared them with the decomposition of ammonia in a helium atmosphere. Studies have shown that calcium oxide can accelerate the decomposition of ammonia in a helium atmosphere, especially the chemical reactions of NH_3 and NO. The typical gas composition of a helium atmosphere and gasifier atmosphere was studied at 900 °C. The calcium oxide in the gasifier atmosphere lost its catalytic activity, the increase in the total pressure could further reduced the ammonia decomposition rate, and the calcium oxide in the gasifier atmosphere enhanced the conversion of NO to NH_3.

In 2001, Monnery et al. [99] studied the high-temperature dissociation and oxidation mechanisms of ammonia at the Claus furnace temperature, and the reaction rate equations for the high-temperature dissociation and oxidation of ammonia were improved with the obtained experimental data. The matching degree of the calculated high-temperature dissociation reaction rates and the experimental data was within 13%, and the matching degree of the calculated ammonia oxidation reaction rates and the experimental data was within 10%.

In 2004, Allison et al. [100] studied a pulsed inductive thruster (PIT) using ammonia as a propellant, performed numerical simulations using a dual-temperature thermochemical model, extended the use range of the original magnetohydrodynamics codes in terms of temperature and density, and compared the results with other models in terms of propellant components and thermodynamic properties.

In 2006, Gianperio Colonna et al. [101] investigated the flow in a nozzle using ammonia as a propellant. A kinetics model of ammonia flow in a supersonic nozzle was established, and the performance of the thruster was found to depend on the degree of ammonia dissociation. For the ammonia dissociation, 2500 K is the cutoff point. At a temperature < 2500 K, the ammonia dissociation is very slow, and the internal states are basically ignored. At temperatures of 3000–5000 K, the dissociation and vibrational excitation energy of ammonia become very important.

In 2006, Bock et al. [102] designed and experimentally validated an ammonia propellant supply system for a 1 kW thermal arcjet thruster and provided a design formula for the flow channel diameter based on a numerical model and experiments in a vacuum, which has important reference significance for the ammonia supply systems of solar thermal thrusters.

Chapter 2
Physical Modeling and Basic Methods

2.1 Introduction

A solar thermal propulsion (STP) system is composed of a concentrator, a thruster and a propellant supply system. To improve the heat exchange efficiency of the system, an integrated structure of regeneratively cooled concentrator and thrust chamber is proposed in this chapter, and the numerical methods required for the simulations are also provided.

2.2 Physical Model of Solar Thermal Thrusters

2.2.1 Model of Solar Concentrator Performance

The concentrator is a main component of an STP system. At present, in China and abroad, a compound parabolic concentrator (CPC) is mostly used as the primary concentrator, a refractive secondary concentrator (RSC) is used as the secondary concentrator, and the temperature of the converging spot is increased by two-stage light concentration. The CPC uses a multiaxis rotating heliostat to reflect sunlight to the concentrator and improve the system's ability to track the sun. The RSC is placed at the focus of the parabola, and the solar radiation concentrated by the concentrator directly heats the absorber/thrust chamber.

The relationship between equilibrium temperature of the solar thermal absorber T_r and the concentration ratio C is [37]:

$$T_r = \left[\frac{(\eta_o - \eta_c)I}{\sigma \varepsilon} \right]^{\frac{1}{4}} C^{\frac{1}{4}} \tag{2.1}$$

© National University of Defense Technology Press 2025
M. Huang et al., *Solar Thermal Thruster*, https://doi.org/10.1007/978-981-97-7490-6_2

The equilibrium temperature of the absorber T_r is a function of incident radiation intensity I, condensation ratio C, optical efficiency η_o, and collection efficiency η_c. Since the characteristics of the radiation energy emitted by the sun and the spatial relationship between the sun and the Earth basically remain unchanged, the solar radiation flux density outside the Earth's atmosphere is generally fixed, that is, the solar radiation intensity $I = 1360$ W/m^2. Optical efficiency η_o is the ratio of the heat absorbed by the absorber to the incident energy, and collection efficiency η_c is the ratio of the absorbed heat of the working fluid to the incident energy. $\sigma = 5.6697 \times 10^{-8}$ W/(m$^2 \cdot$K^4) is the Stephan–Boltzmann constant, and ε is the emissivity of the absorber. The relationship between absorber equilibrium temperature T_r and the concentration ratio C is shown in Fig. 2.1.

The maximum concentration ratio is

$$C_{\max} = \frac{n}{\sin^2 \theta_{\text{sun}}} \tag{2.2}$$

where θ_{sun} is the sun half-acceptance angle $(0.25°)$, and n is the refractive index of the environment.

With the entrance aperture of the absorber as the receiving surface, the concentration ratio at boundary angle of the concentrator being Φ can be expressed as

$$C_{\max} = \frac{\sin^2 \Phi \cos^2(\Phi + \theta_{sun})}{\sin^2 \theta_{sun}} \tag{2.3}$$

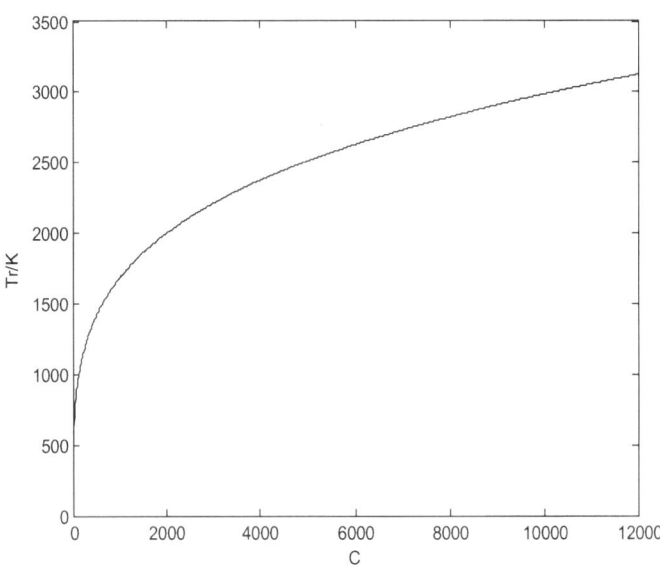

Fig. 2.1 Relationship between the absorber equilibrium temperature and concentration ratio

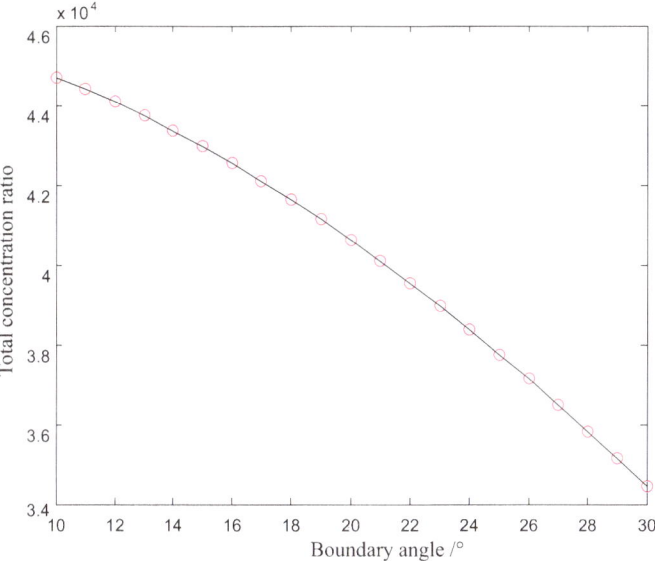

Fig. 2.2 Relationship between total concentration ratio and boundary angle

The boundary angle of the primary concentrator is in a range of 10–30°; thus, the secondary concentration ratio is in the range of 4–36. There is an optimal value that maximizes the concentration ratio after the second concentrator, and the temperature distribution of the secondary concentrator should also be considered. The relationship between the total concentration ratio after the two-stage concentration and the boundary angle is shown in Fig. 2.2.

Therefore, by using the secondary concentrator, a larger concentration ratio can be obtained with a smaller boundary angle. If the obtained concentration ratio is too large (30,000–40,000), the theoretical temperature of the focal spot could be above 5000 K, and the material cannot withstand such a high temperature. In fact, the light could diverge in the absorption cavity, and the exit area of the secondary concentrator is not the area corresponding to the concentration ratio, while the surface area of the absorption cavity is the corresponding area in the calculation of the concentration ratio. Therefore, such high temperatures cannot be reached in practice. Through calculation, the actual concentration ratio is approximately 8000–10,000. In this way, a total concentration ratio that meets the requirements of an STP system can be obtained, and the temperature is within the tolerance of the material. For example, in the STP device developed by the Japan National Laboratory, the theoretical concentration ratio of the concentrator is approximately 8600, and the focal spot temperature obtained in the experiments is 2200 K.

2.2.2   Structure of Integrated Regeneratively Cooled Concentrator and Thrust Chamber

The solar thermal thruster is divided into modules, including an absorber, heat exchanger core, nozzle, and heat insulation layer, and uses radiation heat transfer and convective heat transfer to heat the propellant. The design of the thrust chamber includes optical links to guide sunlight into the thrust chamber and configurations of propellant absorption, energy deposition, and thrust generation. Under the constraints of high-efficiency deposition of working fluid from energy and high-efficiency conversion of working fluid energy to thrust, the optical link is tightly coupled with the configurations of the working fluid injection channel, heating chamber, and thruster, with mutual influences. The configuration design of the high-efficiency heat exchange runner with the laminates and the integrated design of the regeneratively cooled concentrator and the thrust chamber are both important means to improve the heat exchange efficiency of the system.

The regenerative cooling method is used to make the propellant flow in the absorption cavity before entering the high-temperature heat exchange runner, so the RSC is cooled to a certain extent and the propellant is preheated, which improves the use efficiency of solar energy. The integrated design of the RSC and the thrust chamber is shown in Fig. 2.3a. After the propellant enters the absorption cavity, the flow is evenly divided through a porous sleeve so that the propellant can evenly cool the RSC and simultaneously heat the propellant. The propellant flow route is shown in Fig. 2.3b.

Under the designed working conditions, the temperature and pressure variation curves of the propellant flowing through each component region are shown in Fig. 2.4.

The special lateral threaded surface can effectively reduce the reflection of sunlight on the inner wall of the inner cylinder, and the heat absorption efficiency is improved by the spectrally selective absorption coating. Transition metals and semiconductor materials have intrinsic selective solar absorption properties. Among them, HfC has a high absorption value in the solar spectral region and a high melting point; therefore, it can be used to form a solar radiation absorption surface at high temperature. In addition, surface texturing is an effective technique for the selective capture of solar energy. It is accepted that a properly textured surface is rough relative to the solar wavelength and thus can absorb more solar energy. For example, wrinkling the surface into a series of "V" shapes can increase the solar absorptivity to close to 1. Wire meshes, grooves, and electrodeposition coatings on mechanically roughened surfaces, the evaporation of semiconductors under partial vacuum, and surface roughening by sputtering and chemical vapor deposition (CVD) all can texture the surface to enhance sunlight absorption.

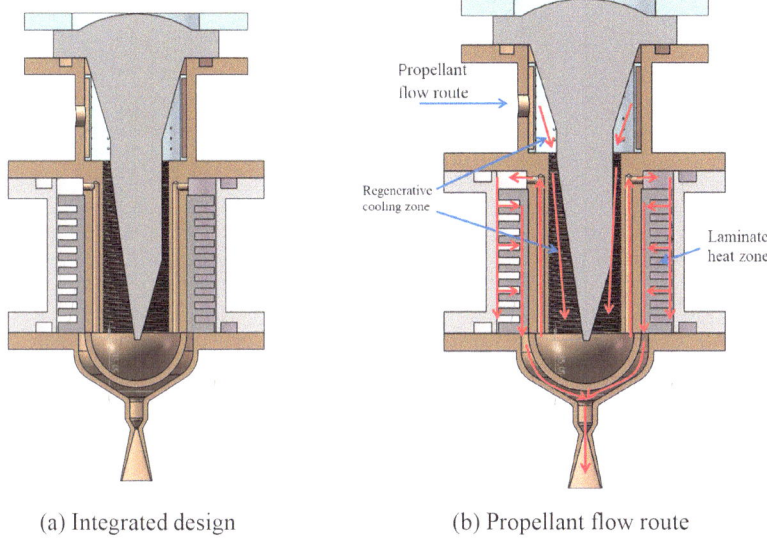

(a) Integrated design   (b) Propellant flow route

Fig. 2.3 Integrated design of the regeneratively cooled RSC and the thrust chamber and the propellant flow route

Fig. 2.4 Temperature and pressure variation curves of propellant flowing through each component region

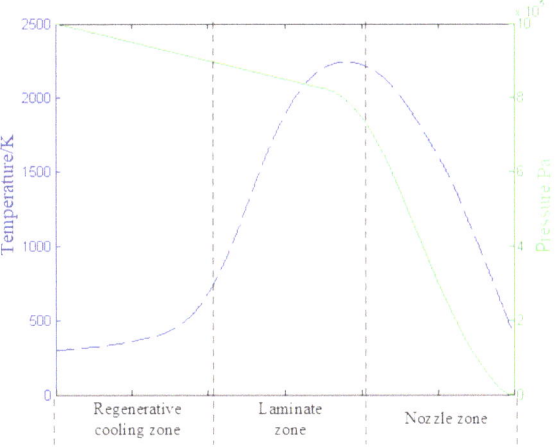

2.2.3 STP System Parameters

The STP system is mainly composed of a sunlight collection and transmission subsystem, an absorber/thrust chamber subsystem, and a propellant supply subsystem, and a design scheme is shown in Fig. 2.5. According to the design requirements, after the sunlight is converged by the two-stage concentrators, the

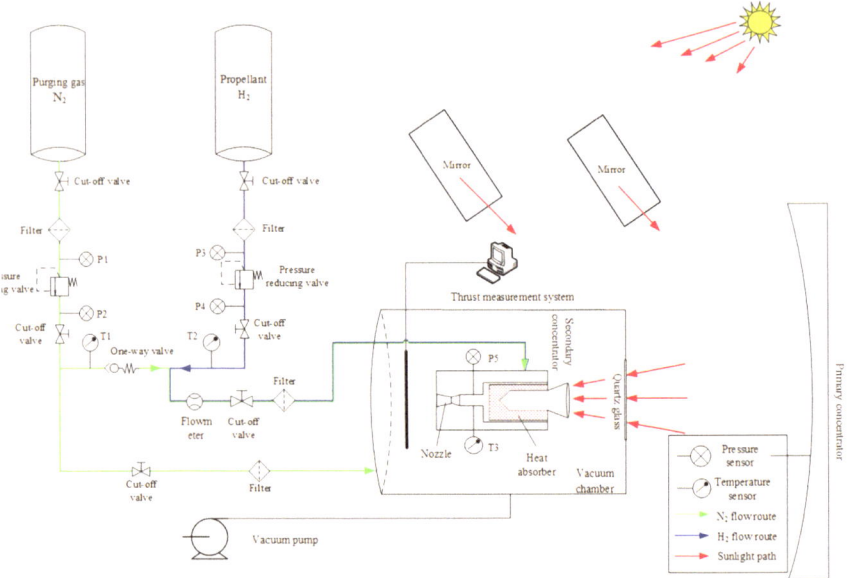

Fig. 2.5 Schematic diagram of the experimental STP system

temperature of the heated hydrogen working fluid must reach 2000 K or higher, the specific impulse should reach 800 s, and the thrust should be no less than 0.5 N. During the ground test, the solar thermal thruster and the secondary concentrator are both placed in the vacuum chamber. After the primary concentration, the sunlight passes through the quartz glass on the vacuum chamber, converges again by the secondary concentrator, and enters the absorption cavity to heat the thruster. The propellant supply system has two routes including purging gas N_2 and propellant H_2.

The performance of the thruster is measured by a specific impulse [103]:

$$I_{sp} = \frac{F}{\dot{m}g} \tag{2.4}$$

where F is the thrust of the thruster and $\dot{m}$ is the mass flow rate of the propellant.

The thrust is defined as:

$$F = \dot{m}u_e + (p_e - p_a)A_e \tag{2.5}$$

where

$$u_e = \sqrt{\frac{2\gamma RT_c}{(\gamma - 1)M}\left[1 - (\frac{p_e}{p_c})^{\frac{\gamma-1}{\gamma}}\right]} \tag{2.6}$$

The specific impulse can be expressed as:

$$I_{sp} = \sqrt{\frac{2\gamma RT_c}{(\gamma - 1)g^2 M}\left[1 - (\frac{p_e}{p_c})^{\frac{\gamma-1}{\gamma}}\right]} \qquad (2.7)$$

When $p_e = p_a$, the thruster power is:

$$P = \frac{1}{2}Fu_e \qquad (2.8)$$

Through the calculation and comparison of thruster parameters under different thrust chamber pressures, the determined thrust chamber parameters are shown in Table 2.1. Under two different solar powers, 5.6 kW and 6.0 kW, the working fluid ammonia temperature in the thrust chamber reaches 2368 K and 2557 K, respectively. The heat transfer efficiency of the former thruster is 89.8%, and the thrust and specific impulse are 0.575 N and 335.3 s, respectively. The heat transfer efficiency of the latter thruster is 88.1%, and the thrust and specific impulse are 0.575 N and 339.2 s, respectively.

Table 2.1 Thruster design parameters

Parameters	Working condition I	Working condition II
Ammonia inlet pressure/ $\times 10^5$ Pa	8	8
Ammonia inlet temperature/K	300	300
Diameter of primary concentrator/m	2.30	2.38
Solar power/kW	**5.6**	**6.0**
Nozzle throat diameter/mm	0.9	0.9
Angle of expansion of nozzle expansion section/($^\circ$)	15	15
Expansion ratio of nozzle expansion section	100	100
Temperature in thrust chamber/K	2368	2557
Thrust/N	**0.575**	**0.575**
Specific impulse/s	**335.3**	**339.2**
Mass flow rate/ $\times 10^{-4}$ kg/	1.75s	1.73
Heat transfer efficiency/(%)	89.8	88.1
Nozzle efficiency/(%)	96.0	96.0
Thruster efficiency/(%)	**86.2**	**84.6**

2.3   Numerical Calculation Methods

2.3.1   *Flow Control Equations*

The basic control equations of fluid motion, the Navier–Stokes equations, are abbreviated as NS equations. In the Cartesian coordinate system, the 3D conservation form of an NS equation can be written as:

$$\frac{\partial Q}{\partial t} + \frac{\partial E}{\partial x} + \frac{\partial F}{\partial y} + \frac{\partial G}{\partial z} = \frac{\partial E_u}{\partial x} + \frac{\partial F_v}{\partial y} + \frac{\partial G_w}{\partial z} \tag{2.9}$$

where

$$Q = \begin{bmatrix} \rho \\ \rho u \\ \rho v \\ \rho w \\ E_t \end{bmatrix} \quad E = \begin{bmatrix} \rho u \\ P + \rho u^2 \\ \rho uv \\ \rho uw \\ (P + E_t)u \end{bmatrix} \quad F = \begin{bmatrix} \rho v \\ \rho uv \\ P + \rho v^2 \\ \rho vw \\ (P + E_t)v \end{bmatrix} \quad G = \begin{bmatrix} \rho w \\ \rho uw \\ \rho vw \\ P + \rho w^2 \\ (P + E_t)w \end{bmatrix}$$

$$E_u = \begin{bmatrix} 0 \\ \tau_{xx} \\ \tau_{xy} \\ \tau_{xz} \\ u\tau_{xx} + v\tau_{xy} + w\tau_{xz} + q_x \end{bmatrix} \quad F_v = \begin{bmatrix} 0 \\ \tau_{xy} \\ \tau_{yy} \\ \tau_{yz} \\ u\tau_{xy} + v\tau_{yy} + w\tau_{yz} + q_y \end{bmatrix}$$

$$G_w = \begin{bmatrix} 0 \\ \tau_{xz} \\ \tau_{yz} \\ \tau_{zz} \\ u\tau_{xz} + v\tau_{yz} + w\tau_{zz} + q_z \end{bmatrix}$$

where u, v, and w represent the velocities in the x, y, and z directions, and P and ρ represent the pressure and density of the fluid, respectively. The total energy of the unit mass of fluid is $E_t = \rho \left[e + \frac{1}{2}(u^2 + v^2 + w^2) \right]$, and e is the specific internal energy of unit mass of fluid. In each vector, the specific expressions of shear stress τ_{ij} and heat transfer term q_i are as follows:

$$\tau_{xx} = \mu \left[-\frac{2}{3}(\nabla \cdot \vec{V}) + 2\frac{\partial u}{\partial x} \right]$$

$$\tau_{yy} = \mu \left[-\frac{2}{3}(\nabla \cdot \vec{V}) + 2\frac{\partial v}{\partial y} \right]$$

$$\tau_{zz} = \mu\left[-\frac{2}{3}(\nabla \cdot \vec{V}) + 2\frac{\partial w}{\partial z}\right]$$

$$\tau_{xy} = \tau_{yx} = \mu\left[\frac{\partial u}{\partial y} + \frac{\partial v}{\partial x}\right]$$

$$\tau_{xz} = \tau_{zx} = \mu\left[\frac{\partial u}{\partial z} + \frac{\partial w}{\partial x}\right]$$

$$\tau_{yz} = \tau_{zy} = \mu\left[\frac{\partial v}{\partial z} + \frac{\partial w}{\partial y}\right]$$

$$\nabla \cdot \vec{V} = \frac{\partial u}{\partial x} + \frac{\partial v}{\partial y} + \frac{\partial w}{\partial z}$$

$$q_x = -\lambda\frac{\partial T}{\partial x},\ q_y = -\lambda\frac{\partial T}{\partial y},\ q_z = -\lambda\frac{\partial T}{\partial z}$$

where λ is the heat transfer coefficient of the fluid and T is the temperature of the fluid.

For a perfect gas, the internal energy is only a linear function of temperature, $e = \frac{1}{\gamma-1} \cdot \frac{P}{\rho}$, where γ is the specific heat ratio of the gas. These equations plus the perfect gas equation $P = \rho R T$ form a closed system of equations.

2.3.2 Turbulence Model

In the turbulent flow region, the turbulent kinetic energy and turbulent flow stress equations, which reflect the effect of the pulsation of turbulent flow on the flow field, can be obtained through the achievable $k - \varepsilon$ equation:

$$\rho\frac{\partial k}{\partial t} = \frac{\partial}{\partial x_i}\left[\left(\mu + \frac{\mu_t}{\sigma_k}\right)\frac{\partial k}{\partial x_i}\right] + G_k + G_b - \rho\varepsilon - Y_M \tag{2.10}$$

$$\rho\frac{\partial \varepsilon}{\partial t} = \frac{\partial}{\partial x_i}\left[\left(\mu + \frac{\mu_t}{\sigma_\varepsilon}\right)\frac{\partial \varepsilon}{\partial x_i}\right] + \rho C_1 S\varepsilon - \rho C_2\frac{\varepsilon^2}{k + \sqrt{v\varepsilon}} + C_{1\varepsilon}\frac{\varepsilon}{k}C_{3\varepsilon}G_b \tag{2.11}$$

where ε is the dissipation rate of turbulent kinetic energy; G_k is the contribution of the average velocity gradient to the turbulent kinetic energy generation term k; G_b is the contribution of buoyancy to the turbulent kinetic energy generation term k, and the effect of gravity is not considered in this book, so $G_b = 0$; Y_M is the contribution of the pulsating expansion of the compressible flow to the total dissipation rate; $C_1 = \max\left|0.43, \frac{\eta}{\eta+5}\right|$, $\eta = Sk/\varepsilon$; and $C_{1\varepsilon}$, C_2, σ_k, and σ_ε are empirical constants, with the values of $C_{1\varepsilon} = 1.44$, $C_2 = 1.9$, and $\sigma_k = 1.0$, $\sigma_\varepsilon = 1.3$.

2.3.3 Heat Transfer Model of Solids

The energy transport equation is used for the solid region, the heat conduction of the solid structure satisfies Fourier law [105], and the differential equation in the steady state is:

$$\frac{\partial^2 T}{\partial x_j \partial x_j} = 0 \tag{2.12}$$

where T is the solid temperature.

The energy balance equation based on the finite element method is expressed as

$$[K]\{T\} = \{Q\} \tag{2.13}$$

where $[K]$ is the transfer matrix, including the thermal conductivity and convective coefficient; $\{T\}$ is the vector of nodal temperature; and $\{Q\}$ is the vector of nodal heat flow rate.

2.3.4 Radiation Heat Transfer Model

When a ray is transmitted in a medium, its energy is gradually attenuated due to the absorption and scattering of the medium. For a beam transmitting along the x direction, the spectral radiation intensity is I_λ. According to Beer's law, the spectral radiation intensity attenuates exponentially along the travelled distance [108]:

$$I_{\lambda,L} = I_{\lambda,0} \exp\left[-\int_0^L \beta_\lambda(x)\mathrm{d}x\right] \tag{2.14}$$

where $I_{\lambda,L}$ is the spectral radiation intensity at $x = L$; $I_{\lambda,0}$ is the spectral radiation intensity at $x = 0$; and β_λ is the spectral attenuation coefficient, in units of 1/m, and is composed of two parts, namely,

$$\beta_\lambda(x) = \kappa_\lambda(x) + \sigma_{s\lambda}(x) \tag{2.15}$$

where κ_λ is the spectral absorption function and $\sigma_{s\lambda}$ is the spectral scattering function.

The RSC is made of a single-crystal sapphire, which absorbs little energy from the solar spectrum. A single-crystal material is theoretically transparent for all solar spectra with wavelengths less than 5 μm; that is, there is no absorption loss; however, a solar spectrum with wavelengths greater than 5 μm could be absorbed by a single-crystal material, causing an energy loss of approximately 0.5%. For this part of the solar spectrum and infrared radiation, the RSC is a non-gray semitransparent material. Therefore, to simulate this radiation heat transfer process, a simplified two-stage spectral band model can be used, i.e., a 5 μm wavelength as the cutoff point of the spectral band.

The RSC medium is at position s, and the radiative transfer equation along the radiative transfer direction h is

$$\frac{dI_\lambda(s, h)}{ds} = -\beta_\lambda(s)I_\lambda(s, h) + S_\lambda(s, h_i) \quad (2.16)$$

where β_λ is the attenuation coefficient, representing the sum of the absorption spectrum and the scattered emission spectrum. S_λ is the radiation source function, which includes the emission source and the scattering sources due to the incidence in all directions in space, and the dominant one is the incident and concentrated sunlight. Then,

$$S_\lambda(s, \mathbf{s}) = \kappa_\lambda(s)I_{b\lambda}(s) + \frac{\sigma_{s\lambda}(s)}{4\pi}$$

$$\int_{\Omega_i=4\pi} I_\lambda(s, \mathbf{s}_i)\Phi_\lambda(\mathbf{s}_i, \mathbf{s})d\Omega_i \quad (2.17)$$

The RSC interface is a selective interface. For a solar spectrum less than 5 μm, the spectral radiation intensity on 2 sides of the interface is equal; for a solar spectrum greater than 5 μm, the radiation intensity at the interface is the sum of two parts: one is the transmitted part of the radiation projected from the environment, and the other is the reflected radiation at the interface on the medium side. Since the interface is a specular surface, then

$$I^+(0, \mu) = \left(\frac{n_m}{n_o}\right)^2 (1 - \rho_o^s)I_o(0, \mu_i) + 2\rho^s I^-(0, -\mu) \quad (2.18)$$

where $\mu_i = \cos\theta_i$, θ_i is the incident angle of the radiation projected from the environment, n is the refractive index, the subscript o represents the outside of the interface, the subscript m represents the inside of the interface, and ρ^s represents the specular reflectance.

The differential equation of unsteady heat transfer between the RSC and the absorber wall is expressed as

$$\rho c\frac{\partial T}{\partial \tau} = \frac{1}{r} \cdot \frac{\partial}{\partial r}\left(\lambda r\frac{\partial T}{\partial r}\right) + \frac{1}{r^2} \cdot \frac{\partial}{\partial \varphi}\left(\lambda\frac{\partial T}{\partial \varphi}\right)$$

$$+ \frac{\partial}{\partial z}\left(\lambda\frac{\partial T}{\partial z}\right) \quad (2.19)$$

where λ is the thermal conductivity, ρ is the density, and c is the specific heat of the material.

In this book, the discrete ordinate (DO) radiation model and the fluid–solid coupling method are used to simulate the radiation heat transfer process and the propellant flow process in the absorber cavity. Because there is medium flow between the radiation surfaces, the RSC is a non-gray semitransparent medium that also participates in the transmission and absorption of solar radiation. In addition, the emission and absorption of solar radiation need to use the non-gray medium model. The DO model can use the gray band model to calculate the radiation of non-gray medium, thus simplifying the calculation; therefore, the DO radiation model is better suited to solve this problem. The DO method, also known as the S_N method, converts the transfer equation (for blackbody or spectrum-based) into a series of partial differential equations, which can be applied in theory with any order and precision.

The DO method is based on the discretization of the directional change in the radiation intensity, and the solution is obtained by solving the radiation transfer equations in a series of discrete directions on the solid angle covering the entire global space (4π).

In a 3D Cartesian coordinate system, using the DO method, the integral term at the right end of Eq. (2.17) is approximately replaced by a numerical integration, and the radiation transfer equation is solved in the discrete direction, that is,

$$
\xi^m \frac{\partial I_k^m}{\partial x} + \eta^m \frac{\partial I_k^m}{\partial y} + \mu^m \frac{\partial I_k^m}{\partial z} = -\beta_k I_k^m + \kappa_k I_{bk}(s)
$$
$$
+ \frac{\sigma_{sk}}{4\pi} \left[\sum_{l=1}^{N\Omega} w^l I_k^l \Phi_k^{m,l} \right] \tag{2.20}
$$

where the values of direction cosine of the radiative transfer direction ξ^m, η^m, μ^m and the integration constant w^l are subject to certain conditions; the superscript l, m represents the discrete lth and mth solid angles along space direction, l, $m = 1, 2, \ldots, N\Omega$; $N\Omega$ is the total number of solid angles that are discrete in the 4π space direction; and $\Phi_k^{m,l} = \Phi_k(\Omega^m, \Omega^l)$ is the discretized scattering phase function.

For the opaque boundary wall surface with diffuse emission and diffuse reflection (the subscript w represents the wall surface), the corresponding boundary condition is

$$
I_{k,w}(s) = \varepsilon_{k,w} I_{bk,w}
$$
$$
+ \frac{1 - \varepsilon_{k,w}}{\pi} \int_{\mathbf{n}_w \cdot \mathbf{s}_i < 0} I_{k,w}(s_i) |\mathbf{n}_w \cdot \mathbf{s}_i| d\Omega_i \tag{2.21}
$$

where $\varepsilon_{k,w}$ is the wall band emissivity and $\mathbf{n}$ is the wall normal vector.

If the refractive index of the medium $n_k = n = 1$, Eq. (2.21) is discretized to obtain

$$
I_{k,w}^m = \varepsilon_{k,w} \frac{\sigma B_{k,T_w} T_w^4}{\pi}
$$

$$+ \frac{(1 - \varepsilon_{k,w})}{\pi} \sum_{n_w \cdot s_l < 0} w^l I_{k,w}^l |n_w \cdot s_l| \quad (n_w \cdot s_m > 0) \tag{2.22}$$

$$B_{k,T_w} = \int_{\Delta \lambda_k} E_{b\lambda}(T_w) d\lambda \Big/ \left[\int_0^\infty E_{b\lambda}(T_w) d\lambda \right] \tag{2.23}$$

where B_{k,T_w} is the proportion of radiation energy in region k of the spectral band model in the total radiation energy at the wall temperature T_w.

As shown in Fig. 2.6, the direction vector r^m is used to define the center of each solid angle, the subscripts E, W, S, N, T, and B represent the center nodes of each control body adjacent to the control body P, and the subscripts e, w, s, n, t, and b represent the boundaries of control body P, so the integral equation on the control body can be expressed as

$$\xi^m A_x (I_{k,e}^m - I_{k,w}^m) + \eta^m A_y (I_{k,n}^m - I_{k,s}^m) + \mu^m A_z (I_{k,t}^m - I_{k,b}^m)$$
$$= -\beta_k I_{k,P}^m V_P + \kappa_k I_{bk,P} V_P + \frac{\sigma_{sk}}{4\pi} \left[\sum_{l=1}^{N\Omega} w^l I_{k,P}^l \Phi_k^{m,l} \right] V_P \tag{2.24}$$

where V_P is the control body volume, and $V_P = A_x A_y A_z$.

In radiation heat transfer, the nth order moment is generally defined as

$$\int_{\Omega = 4\pi} f(s) s^n d\Omega \tag{2.25}$$

Then, the zero-order moment and the first-order moment can be expressed as

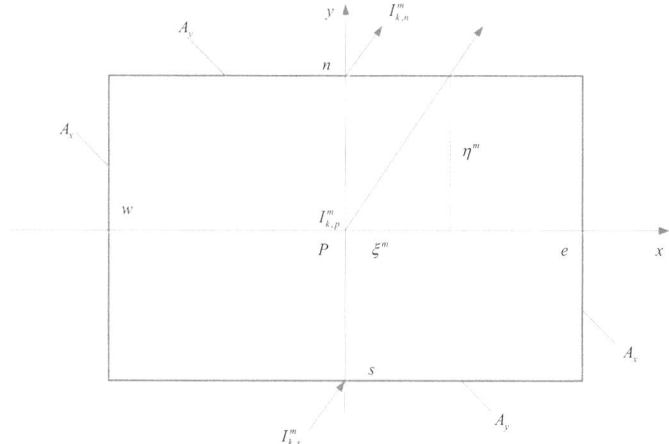

Fig. 2.6 Calculation model of the DO method

$$\int_{\Omega=4\pi} f(s)\,d\Omega \quad \int_{\Omega=4\pi} f(s)s\,d\Omega \tag{2.26}$$

Let $f(s) = I(s)$; then, the zero-order moment and first-order moment can be expressed as the projected radiation H and radiation heat flux q, respectively, as

$$H = \int_{\Omega=4\pi} I(s)\,d\Omega \quad q = \int_{\Omega=4\pi} I(s)s\,d\Omega \tag{2.27}$$

The corresponding discrete equations are

$$H = \sum_{l=1}^{N\Omega} w^l I^l \quad q = \sum_{l=1}^{N\Omega} w^l I^l s^l \tag{2.28}$$

The discrete equation of the net heat flux of wall radiation is

$$
\begin{aligned}
q_w^r &= \int_0^\infty \varepsilon_{\lambda,w} \left[\pi I_{b\lambda,w} - \int_{n_w \cdot s_i < 0} I_{\lambda,w}(s_i) |n_w \cdot s_i| \, d\Omega_i \right] d\lambda \\
&= \sum_{k=1}^{M_b} \varepsilon_{k,w} \left(E_{bk,w} - \sum_{n_w \cdot s_i < 0} w^l I_{k,w}^l |n_w \cdot s_l| \right) \\
&= \sum_{k=1}^{M_b} \varepsilon_{k,w} \left(B_{k,T_w} \sigma T_w^4 - \sum_{l=1}^{N\Omega/2} w^l I_{k,w}^l |n_w \cdot s_l| \right) \tag{2.29}
\end{aligned}
$$

where M_b is the total number of spectral bands divided by the radiation characteristics versus wavelength in the spectral band approximation method.

2.3.5 Thermal Stress Model of RSC

Due to the uneven heating of RSCs, thermal stress can be generated inside the RSCs. If the thermal stress is too large, the RSCs can crack. The stress should satisfy the following coordination equation [111]:

$$\begin{cases} \dfrac{\partial^2 \varepsilon_{xx}}{\partial y^2} + \dfrac{\partial^2 \varepsilon_{yy}}{\partial x^2} = 2\dfrac{\partial^2 \varepsilon_{xy}}{\partial x \partial y} \\[2mm] 2\dfrac{\partial^2 \varepsilon_{xx}}{\partial y \partial z} = \dfrac{\partial}{\partial x}\left\{ -\dfrac{\partial \varepsilon_{yz}}{\partial x} + \dfrac{\partial \varepsilon_{zx}}{\partial y} + \dfrac{\partial \varepsilon_{xy}}{\partial z} \right\} \\[2mm] \dfrac{\partial^2 \varepsilon_{yy}}{\partial z^2} + \dfrac{\partial^2 \varepsilon_{zz}}{\partial y^2} = 2\dfrac{\partial^2 \varepsilon_{yz}}{\partial y \partial z} \\[2mm] 2\dfrac{\partial^2 \varepsilon_{yy}}{\partial z \partial x} = \dfrac{\partial}{\partial y}\left\{ -\dfrac{\partial \varepsilon_{zx}}{\partial y} + \dfrac{\partial \varepsilon_{xy}}{\partial z} + \dfrac{\partial \varepsilon_{yz}}{\partial x} \right\} \\[2mm] \dfrac{\partial^2 \varepsilon_{zz}}{\partial x^2} + \dfrac{\partial^2 \varepsilon_{xx}}{\partial z^2} = 2\dfrac{\partial^2 \varepsilon_{zx}}{\partial z \partial x} \\[2mm] 2\dfrac{\partial^2 \varepsilon_{zz}}{\partial x \partial y} = \dfrac{\partial}{\partial z}\left\{ -\dfrac{\partial \varepsilon_{xy}}{\partial z} + \dfrac{\partial \varepsilon_{yz}}{\partial x} + \dfrac{\partial \varepsilon_{zx}}{\partial y} \right\} \end{cases} \tag{2.30}$$

The stress component acting on each surface should satisfy the following equilibrium differential equation:

$$\begin{cases} \dfrac{\partial \sigma_{xx}}{\partial x} + \dfrac{\partial \sigma_{yx}}{\partial y} + \dfrac{\partial \sigma_{zx}}{\partial z} + X = \rho\dfrac{\partial^2 u_x}{\partial t^2} \\[2mm] \dfrac{\partial \sigma_{xy}}{\partial x} + \dfrac{\partial \sigma_{yy}}{\partial y} + \dfrac{\partial \sigma_{zy}}{\partial z} + Y = \rho\dfrac{\partial^2 u_y}{\partial t^2} \\[2mm] \dfrac{\partial \sigma_{xz}}{\partial x} + \dfrac{\partial \sigma_{yz}}{\partial y} + \dfrac{\partial \sigma_{zz}}{\partial z} + Z = \rho\dfrac{\partial^2 u_z}{\partial t^2} \end{cases} \tag{2.31}$$

where σ_{xx}, σ_{yx}, and σ_{zx} are the stress components on the microelement area $dydz$; σ_{yx}, σ_{yy}, and σ_{yz} are the stress components on the microelement area $dzdx$; σ_{zx}, σ_{zy}, and σ_{zz} are the stress components on the microelement area $dxdy$; u_x, u_y, and u_z are the displacement components at a certain point during deformation; and X, Y, and Z are the volume force components.

Strain is the sum of two parts: one part is due to temperature change, and the other part is due to stress. According to Hooke's law,

$$
\left\{
\begin{aligned}
\varepsilon_{xx} &= \frac{\partial u_x}{\partial x} = \frac{1}{E}\{\sigma_{xx} - \nu(\sigma_{yy} + \sigma_{zz})\} \\
&\quad + \alpha\tau = \frac{1}{2G}\left(\sigma_{xx} - \frac{\nu}{1+\nu}\Theta_s\right) + \alpha\tau \\
\varepsilon_{yy} &= \frac{\partial u_y}{\partial y} = \frac{1}{E}\{\sigma_{yy} - \nu(\sigma_{xx} + \sigma_{zz})\} \\
&\quad + \alpha\tau = \frac{1}{2G}\left(\sigma_{yy} - \frac{\nu}{1+\nu}\Theta_s\right) + \alpha\tau \\
\varepsilon_{zz} &= \frac{\partial u_z}{\partial z} = \frac{1}{E}\{\sigma_{zz} - \nu(\sigma_{zz} + \sigma_{yy})\} \\
&\quad + \alpha\tau = \frac{1}{2G}\left(\sigma_{zz} - \frac{\nu}{1+\nu}\Theta_s\right) + \alpha\tau
\end{aligned}
\right.
\tag{2.32}
$$

$$
\varepsilon_{xy} = \frac{\sigma_{xy}}{2G}, \quad \varepsilon_{yz} = \frac{\sigma_{yz}}{2G}, \quad \varepsilon_{zx} = \frac{\sigma_{zx}}{2G}
\tag{2.33}
$$

where $\Theta_s = \sigma_{xx} + \sigma_{yy} + \sigma_{zz}$, E is the longitudinal elastic modulus of the material, G is the shear modulus of elasticity, ν is the Poisson's ratio, and the relationship between elasticity coefficients is $2G = E/(1+\nu)$.

The stress coordination equations are

$$
\left\{
\begin{aligned}
\Delta\sigma_{xx} &+ \frac{1}{1+\nu}\cdot\frac{\partial^2\Theta_s}{\partial x^2} = -\alpha E\left(\frac{1}{1-\nu}\Delta\tau + \frac{1}{1+\nu}\cdot\frac{\partial^2\tau}{\partial x^2}\right) \\
&- \left\{\frac{\nu}{1-\nu}\left(\frac{\partial X}{\partial x} + \frac{\partial Y}{\partial y} + \frac{\partial Z}{\partial z}\right) + 2\frac{\partial X}{\partial x}\right\} \\
\Delta\sigma_{yy} &+ \frac{1}{1+\nu}\cdot\frac{\partial^2\Theta_s}{\partial y^2} = -\alpha E\left(\frac{1}{1-\nu}\Delta\tau + \frac{1}{1+\nu}\cdot\frac{\partial^2\tau}{\partial y^2}\right) \\
&- \left\{\frac{\nu}{1-\nu}\left(\frac{\partial X}{\partial x} + \frac{\partial Y}{\partial y} + \frac{\partial Z}{\partial z}\right) + 2\frac{\partial Y}{\partial y}\right\} \\
\Delta\sigma_{zz} &+ \frac{1}{1+\nu}\cdot\frac{\partial^2\Theta_s}{\partial z^2} = -\alpha E\left(\frac{1}{1-\nu}\Delta\tau + \frac{1}{1+\nu}\cdot\frac{\partial^2\tau}{\partial z^2}\right) \\
&- \left\{\frac{\nu}{1-\nu}\left(\frac{\partial X}{\partial x} + \frac{\partial Y}{\partial y} + \frac{\partial Z}{\partial z}\right) + 2\frac{\partial Z}{\partial z}\right\} \\
\Delta\sigma_{xy} &+ \frac{1}{1+\nu}\cdot\frac{\partial^2\Theta_s}{\partial x\partial y} = -\frac{\alpha E}{1+\nu}\cdot\frac{\partial^2\tau}{\partial x\partial y} - \left(\frac{\partial X}{\partial y} + \frac{\partial Y}{\partial x}\right) \\
\Delta\sigma_{yz} &+ \frac{1}{1+\nu}\cdot\frac{\partial^2\Theta_s}{\partial y\partial z} = -\frac{\alpha E}{1+\nu}\cdot\frac{\partial^2\tau}{\partial y\partial z} - \left(\frac{\partial Y}{\partial z} + \frac{\partial Z}{\partial y}\right) \\
\Delta\sigma_{zx} &+ \frac{1}{1+\nu}\cdot\frac{\partial^2\Theta_s}{\partial z\partial x} = -\frac{\alpha E}{1+\nu}\cdot\frac{\partial^2\tau}{\partial z\partial x} - \left(\frac{\partial Z}{\partial x} + \frac{\partial X}{\partial z}\right)
\end{aligned}
\right.
\tag{2.34}
$$

2.3.6 *Chemical Kinetics Model*

For the chemical reaction of a multicomponent gas mixture, the general expression for the elementary reaction in the system can be expressed as [111]:

$$\sum_{i=1}^{n} v_i' Z_i \overset{k_+}{\underset{k_-}{\longleftrightarrow}} \sum_{i=1}^{n} v_i'' Z_i \tag{2.35}$$

where Z_i represents any component in the system, v_i' and v_i'' are the stoichiometric coefficients (or stoichiometric moles) of the reactant and the product, respectively, and k_+ and k_- represent the rate constants for the forward reaction and reverse reaction, respectively. The forward and reverse reaction rate equations are as follows:

$$\frac{d[Z_i]}{dt} = (v_i'' - v_i')k_+ \prod_i [Z_i]^{v_i'} \tag{2.36}$$

and

$$\frac{d[Z_i]}{dt} = -(v_i'' - v_i')k_- \prod_i [Z_i]^{v_i''} \tag{2.37}$$

Therefore, the net production rate of any component is

$$\frac{d[Z_i]}{dt} = (v_i'' - v_i')(k_+ \prod_i [Z_i]^{v_i'} - k_- \prod_i [Z_i]^{v_i''}) \tag{2.38}$$

The equilibrium constant is defined as

$$K_{eq} = \frac{k_+}{k_-} = \prod_i [Z_i]^{*v_i''} / \prod_i [Z_i]^{*v_i'} \tag{2.39}$$

The reaction rate is calculated by the Arrhenius equation:

$$k = AT^\delta e^{-E_\alpha/RT} \tag{2.40}$$

where A is a preexponential factor, E_α is the activation energy, and δ is the temperature index.

The reverse process of the reaction can be calculated from the principle of chemical equilibrium:

$$K_r = \frac{K_d}{K_{eq}} \tag{2.41}$$

where K_d is the forward reaction rate and K_{eq} is a chemical equilibrium constant, which are determined by

$$\ln(K_{eq}) = K_\infty + K_p \left(\frac{1000}{T} \right)^{q_p} + K_e e^{-\frac{T}{q_e}} \tag{2.42}$$

2.4 Boundary Conditions

The flow boundary conditions: The propellant inlet pressure is 0.8 MPa, the inlet temperature is 300 K, and vacuum is at the nozzle exit.

Temperature boundary conditions: The STP system has a high specific impulse. When hydrogen is used as the working fluid, the specific impulse can exceed 800 s, the corresponding propellant temperature can reach above 2300 K after heating, and the solar radiation power on the inner wall of the heat exchanger core is 1.2×10^6 W/ m^2. On the low-temperature side, the thermal insulation on the outside of the thruster is considered, so an adiabatic condition is set. The focus is placed on the fluid–solid coupling effect of the laminates.

At the fluid–solid coupling interface, the standard wall function is used to treat the flow boundary layer and thermal boundary layer. The flow velocity and temperature at the first internal node P parallel to the wall should satisfy the logarithmic distribution law, and the equivalent viscosity coefficient and equivalent thermal conductivity between node P and the wall are obtained:

$$\mu_t = \left[\frac{y_P(c_\mu^{1/4} k_P^{1/2})}{\nu} \right] \frac{\mu}{\ln(Ey_P^+)/\kappa} \tag{2.43}$$

$$\lambda_t = \frac{y_P^+ \mu c_P}{\sigma_T [\ln(Ey_P^+)/\kappa + P]} \tag{2.44}$$

where the von Karman constant $\kappa = 0.4 - 0.42$, a constant $c_\mu = \frac{1}{A_0 + A_S U^* \frac{k}{\varepsilon}}$, σ_T is the turbulent Prandtl number, and σ_L is the molecular Prandtl number, $P = 9\left(\frac{\sigma_L}{\sigma_T} - 1 \right)\left(\frac{\sigma_L}{\sigma_T} \right)^{-1/4}$.

From this, the shear stress and heat flux at the wall can be calculated.

The value of k_P on interior node P can still be calculated according to the equation of turbulent kinetic energy k, and the boundary condition is taken as $\left(\frac{\partial k}{\partial y} \right)_W = 0$, where y is the coordinate perpendicular to the wall. After k_P is known, ε_P can be obtained.

$$\varepsilon_P = \frac{c_\mu^{3/4} k_P^{3/2}}{\kappa y_P} \tag{2.45}$$

Chapter 3
Model and Parameter Calculation for a Solar Thermal Thruster operating under Variable Working Conditions

3.1 Introduction

The main advantage of a solar thermal thruster system operating under variable working conditions is that the thruster can work under different conditions and provide different thrusts, which shortens the response time of orbital maneuver and attitude control and has the advantages of a clean energy source, high reliability, and multiple reuses.

To meet the requirements for variable thrust, a solar thermal thruster system operating under variable working conditions can change the thrust by changing the incident solar energy and the flow rate of the propellant/working fluid. This chapter mainly introduces the research object—a solar thermal thruster model, and calculates and analyzes the relevant performance parameters of the thruster according to mission requirements.

3.2 Energy Conversion Mechanism for a Solar Thermal Thruster

A schematic diagram of a solar thermal thruster system operating under variable working conditions is shown in Fig. 3.1. The propellant/working fluid is heated by indirect heating. The sunlight is focused by the primary and secondary concentrators and then radiates on the heat exchanger core to raise the temperature of the inner wall. The heat is transferred to the propellant/working fluid through a laminated heat exchanger core in the forms of heat conduction and convection, and as the temperature of the propellant/working fluid increases, the working fluid is forced through the nozzle, converting the internal energy of the working fluid into kinetic energy to generate thrust.

© National University of Defense Technology Press 2025
M. Huang et al., *Solar Thermal Thruster*, https://doi.org/10.1007/978-981-97-7490-6_3

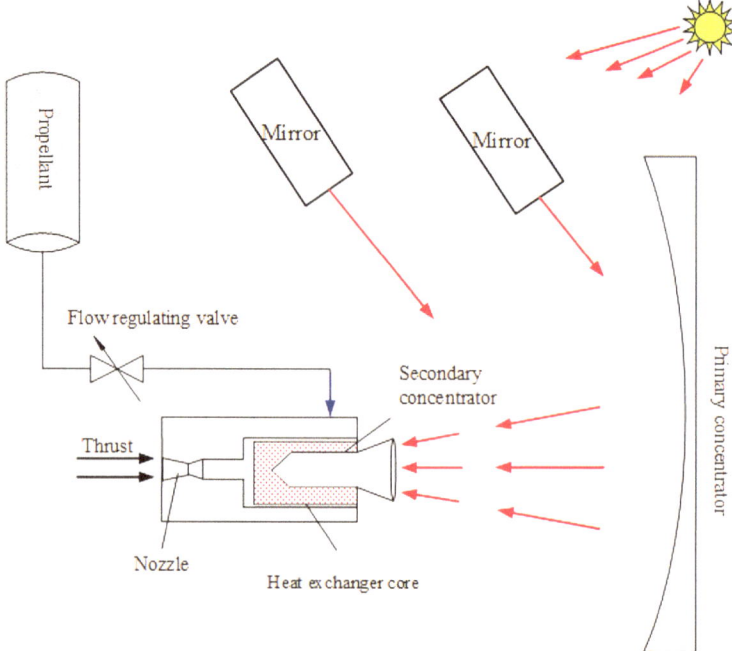

Fig. 3.1 Schematic diagram of the principle of solar thermal propulsion (STP)

The thruster concentrator system mainly includes primary concentrators and refractive secondary concentrators (RSCs) that focus low-density solar energy into high-density solar energy. The propellant supply system mainly consists of propellant tanks, related pipelines and valves that store and supply propellant. The heat absorber/thrust chamber system mainly consists of a laminated heat exchanger core and a nozzle; the core absorbs the solar energy gathered by the secondary concentrator and then heats the propellant/working fluid, while the nozzle converts the thermal energy of the working fluid into mechanical energy to generate thrust.

3.3 Solar Thermal Thruster Operating Under Variable Working Conditions

The overall structure of a solar thermal thruster operating under variable working conditions is shown in Fig. 3.2. The thruster design uses regenerative cooling technology to cool the secondary concentrator, and the proposed design uses the laminated heat exchanger to perform high-efficiency heat exchange for the propellant/working fluid. Figure 3.3 shows the flow route of the propellant in the thruster. After the propellant enters the thruster from the inlet, it flows through a porous sleeve

Fig. 3.2 Overall structure of
a solar thermal thruster
operating under variable
working conditions

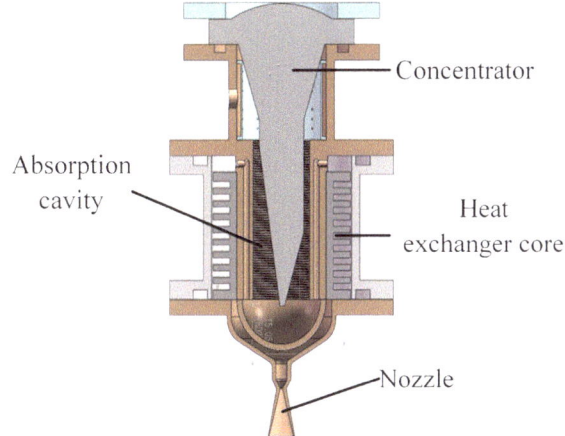

to evenly distribute the propellant around the concentrator and then sequentially
flows through the upper part and lower energy output part of the secondary concentrator. During this process, the propellant absorbs heat from the concentrator through
convective heat transfer to prevent the temperature from becoming too high; i.e., the
propellant is preheated, further heated to a high temperature by flowing through a
runner through a heat exchanger core, and then ejected through a nozzle to generate
thrust.

3.4   Calculation of the Solar Radiation Area

Of the total solar energy absorbed by the thruster, a part is used to heat the working
gas and is converted to the kinetic energy of the working gas, while the other part
is consumed. According to the law of energy conservation, the following can be
obtained:

$$Q = \eta \gamma S \phi_{\text{sol}} = \frac{1}{2} \dot{m} u_e^2 \tag{3.1}$$

where γ is the optical efficiency of the concentrator, i.e., the proportion of the energy
absorbed by the concentrator from the total incident energy; η is the efficiency of
the thruster; S is the area of the concentrator, in m^2; Q is the energy needed by the
thruster per unit time, in W; ϕ_{sol} is the solar constant, and $\phi_{\text{sol}} = 1353 \text{W/m}^2$ is used
as the standard value in calculations.

If the thruster is working normally within Δt and the thrust force F is constant at
this time, then the total impulse of the engine during Δt is:

$$I = F \cdot \Delta t \tag{3.2}$$

Fig. 3.3 Propellant flow
route inside the thruster

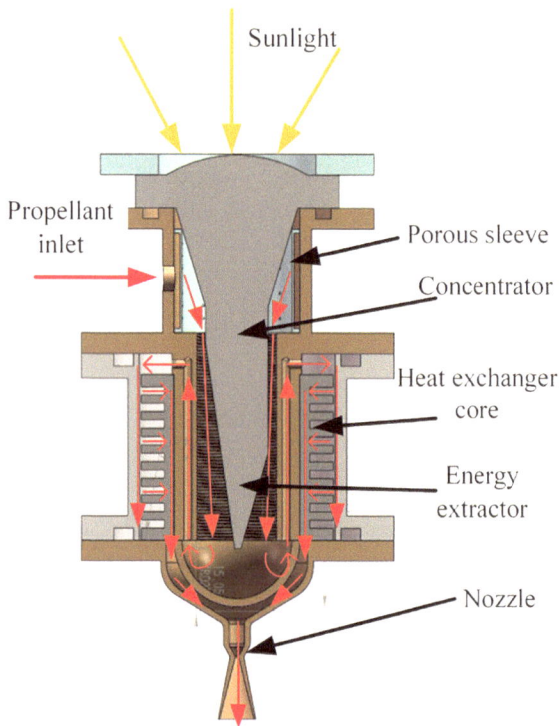

The specific impulse I_{sp} is defined as:

$$I_{sp} = \frac{I}{mg_0} \tag{3.3}$$

The effective exhaust speed u_e is expressed as follows:

$$u_e = I_{sp}g_0 \tag{3.4}$$

Then,

$$\frac{F}{S} = \frac{2\gamma\eta\phi_{sol}}{g_0I_{sp}} \tag{3.5}$$

The estimated optical efficiency of the concentrator $\gamma = 0.5$, the thruster specific impulse $I_{sp} = 340s$ (when ammonia is used as the propellant/working fluid), and the designed thruster efficiency $\eta = 0.8$. After substituting into Equation (3.5), $\frac{F}{S} = 0.4\text{N/m}^2$.

Table 3.1 Corresponding radiation area calculation results under each working condition

Working condition/N	0.1	0.2	0.3	0.4	0.5
Radiation area/m^2	0.25	0.5	0.75	1.0	1.25

The thrusts under each designed working condition are calculated. Table 3.1 shows the calculation results of the solar thermal thruster system operating under variable working conditions.

3.5 Concentration Ratio for a Solar Thermal Thruster Operating Under Variable Working Conditions

Solar energy is a clean energy source with a low energy density. To condense low-density solar energy into high-density solar energy, sunlight is concentrated mainly by solar concentrators. The concentrator is the basis for the normal operation of a solar thermal thruster under variable working conditions. This concentration technology is a core component for solar thermal thruster applications, and the quality of the concentration system directly determines the performance of a solar thermal thruster operating under variable working conditions.

To provide more effective solar energy concentration and collection, the solar energy in a solar thermal thruster concentrator system is focused and extracted mainly by increasing the number of concentrators. At present, a double concentration method is typically used that involves a primary concentrator and the secondary concentrator. The specific process of energy conversion in the thruster is as follows: first, the primary concentration of sunlight is completed by the primary concentrator; second, the solar energy derived after primary convergence is transferred to the secondary concentrator; third, the solar energy is further focused through the reflection and refraction of the sunlight inside the secondary concentrator and transferred to the energy extractor of the secondary concentrator; fourth, the solar energy is transferred to the laminated heat exchanger core through heat transfer methods, such as convective heat transfer and radiation. The above analysis shows that the roles of the primary and secondary concentrators are different. The purpose of the primary concentrator is to achieve a high concentration ratio and convert low-density solar energy into high-density solar energy, while the secondary concentrator achieves energy transfer and extraction and outputs high-density solar energy.

3.5.1 Primary Concentrator for a Solar Thermal Thruster

Based on the shape and performance characteristics of the primary concentrator, typical solar concentrators and performance parameters are shown in Table 3.2.

Table 3.2 Concentration ratios and operating temperature ranges of typical concentrators

Type	Typical concentration ratio range	Working temperature/K
Spherical condenser	50–150	573–873
Fresnel lens	100–1000	573–1273
Compound parabolic concentrator (CPC)	500–3000	773–2273
Tower concentrator	1000–3000	773–2273
Parabolic concentrator	15–50	473–573

The working temperature of the solar thermal thruster ranges from 1000–2500 K and can sometimes exceed 3000 K. The CPC meets the temperature requirements of the thruster, with a large concentration ratio. At the same time, the CPC has a simple principle, convenient manufacturing process, low cost, and better solar concentration effect; therefore, it is an ideal concentrator for a solar thermal thruster operating under variable working conditions.

3.5.2   Secondary Concentrator for a Solar Thermal Thruster

The energy density of sunlight after primary concentration is far below the normal working requirements of solar thermal thrusters, and the total concentration ratio is low. Therefore, it is necessary to add a secondary concentrator to the system to increase the total concentration ratio and improve the solar energy density.

The RSC is composed of a spherical surface, a conical structure and an energy extractor, as shown in Fig. 3.4. The light path propagation and energy transfer are performed by the reflection and total emission of the light inside the secondary concentrator. As the core component for heat exchange cavity, the secondary concentrator is mainly composed of three parts: a concentrator, which mainly realizes the collection and total reflection of the sunlight; an energy extractor, which transmits the solar energy collected by the secondary concentrator to the outside to heat the laminated heat exchanger core; and a fixation component, which can fix the secondary concentrator during the installation process.

3.5.3   Total Concentration Ratio for a Concentration System

During the placement process of the two concentrators, the optical axes of the RSC and the parabolic primary concentrator need to be aligned to achieve high-efficiency light concentration, and the center of the light path entrance of the RSC must be at the focal point of the primary concentrator. Through energy convergence, the solar energy

Fig. 3.4 3D stereogram of RSC

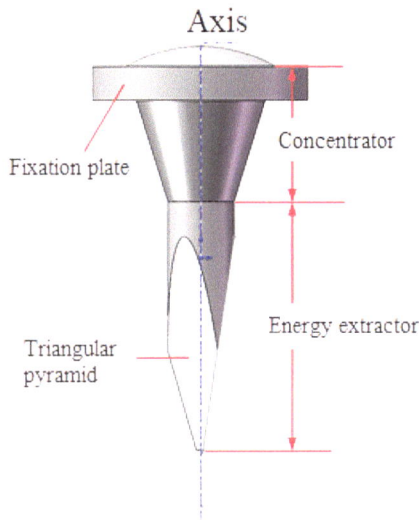

can directly heat the heat exchanger core [113]. The relationship between the equilibrium temperature T_r of the solar energy absorber and the geometric concentration ratio C can be obtained from existing literature [114]:

$$T_r = \left[\frac{(\eta_o - \eta_c)\phi_{sol}}{\sigma \varepsilon} \right]^{\frac{1}{4}} C^{\frac{1}{4}} \tag{3.6}$$

The equilibrium temperature T_r of the absorber is a function of the incident radiation intensity ϕ_{sol}, the total concentration ratio C, the optical efficiency η_o and the collection efficiency η_c. Figure 3.5 shows the relationship between the equilibrium temperature T_r of the solar energy absorber and the geometric concentration ratio C.

For the normal operation of a solar thermal thruster, the temperature of the absorption cavity needs to be increased to above 2000 K. As shown in Fig. 3.5, when the absorption cavity temperature reaches 2000 K, the total concentration ratio needs to exceed 2000, and when the absorption cavity temperature reaches 2400 K, the total concentration ratio needs to exceed 5000.

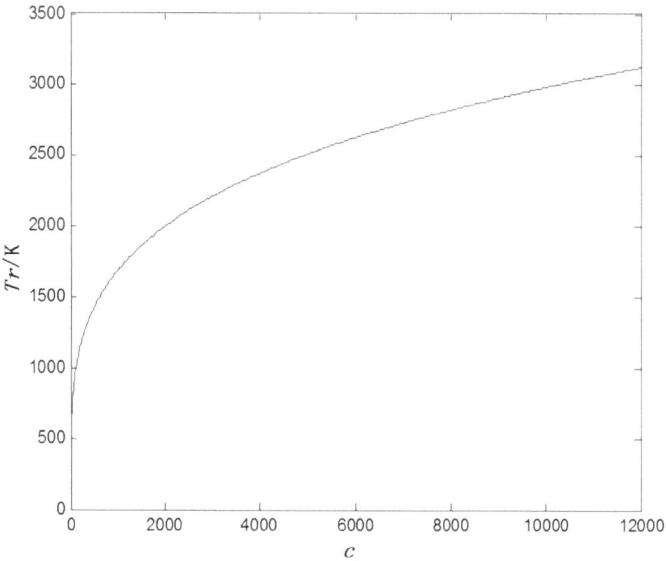

Fig. 3.5 Relationship of the cavity temperature and concentration ratio of the concentrator

3.6 Thrust Change Methods for Thrusters Operating Under Variable Working Conditions

The methods for changing thrust include changing the propellant flow rate, changing the solar energy, and changing the cross-sectional area of the nozzle throat. However, when considering the machining difficulty of microthrust nozzles, changing the thrust when operating under variable working conditions mainly involves changing the solar energy or the propellant flow rate. When calculating this change, the following assumptions are made:

1. The working gas in the thruster is assumed to be an ideal gas.
2. There is no heat transfer across the thrust chamber wall, so the flow is adiabatic.
3. There is no shock wave or discontinuity in the nozzle flow, and the flow is steady.
4. There is no significant friction inside the nozzle, so all boundary layer effects are ignored.
5. The working fluid speed at the thruster nozzle entrance is very low and approximately zero.
6. The exhaust speed at the thruster nozzle is solely axial speed.

For the parameter calculation of the solar thermal thrust chamber, the temperature of hydrogen (1300 K) is selected as the reference temperature. The relevant physical parameters of hydrogen gas are as follows: molar mass $M = 2.016 \times 10^{-3}$ kg/mol, specific heat ratio $\gamma = 1.404$, density $\rho = 0.0189$ kg/m^3, average specific heat capacity at constant pressure $C_p = 1.56 \times 10^4$ J/(kg · K), dynamic viscosity $\mu = 24.08 \times 10^{-6}$ kg/(m · s), and average thermal conductivity $k = 0.568$ W/(m · K).

Since the flow in the nozzle is an adiabatic process without work, there is:

$$h_0 = h + \frac{u^2}{2} = \textbf{constant} \tag{3.7}$$

From the conservation of energy, we obtain:

$$h_c - h_e = \frac{1}{2}\left(u_e^2 - u_c^2\right) = C_p(T_c - T_e) \tag{3.8}$$

The nozzle exhaust speed can be expressed as:

$$u_e = \sqrt{\frac{2k}{k-1}RT_c\left[1 - \left(\frac{P_e}{P_c}\right)^{\frac{k-1}{k}}\right] + u_c^2} \tag{3.9}$$

The mass flow rate of the nozzle is:

$$\dot{m} = \frac{P_c}{RT_c}\left(\frac{2}{k+1}\right)^{\frac{1}{k-1}}\sqrt{\frac{2k}{k+1}RT_c}\cdot A_t \tag{3.10}$$

The vacuum thrust of the engine is:

$$F = \dot{m}u_e + A_e P_e \tag{3.11}$$

The system propulsion efficiency is defined as:

$$\eta = \frac{\dot{m}u_e^2}{2(\dot{m}h_i + P)} \tag{3.12}$$

It is assumed that the propulsion system efficiency is designed as $\eta = 0.8$, P is the absorbed solar energy, and h_i is the inlet enthalpy, which is a single-valued function of temperature.

Therefore, the obtained equations are as follows:

$$\begin{cases} \eta = \dfrac{\dot{m}_2 u_e^2}{2(\dot{m}h_i + P)} \\[2mm] F = \dot{m}_2 u_e \\[2mm] u_e = \sqrt{\dfrac{2k}{k-1}RT_c} \\[2mm] \dot{m} = \dfrac{P_c}{RT_c}\left(\dfrac{2}{k+1}\right)^{\frac{1}{k-1}}\sqrt{\dfrac{2k}{k+1}RT_c}\cdot A_t \end{cases} \tag{3.13}$$

Table 3.3 Parameters under each working condition when solar energy remains unchanged

Working condition	0.1 N	0.2 N	0.3 N	0.4 N	0.5 N
Solar energy (W)	480	480	480	480	480
Mass flow rate ($\times 10^5$ kg/s)	1.178	3.821	7.004	10.396	13.886
Exhaust speed (m/s)	8488.5	5233.8	4283.5	3847.7	3600
Total temperature (K)	2000	760	509	411	360
Overall pressure (MPa)	0.2	0.4	0.6	0.8	1.0

3.6.1   Changing the Propellant Flow Rate While Maintaining the Solar Energy

The theoretical temperature of the propellant/working fluid can exceed 2000 K after passing through the heat exchanger core. When hydrogen is used as the working fluid, the designed thrust chamber temperature and pressure are $T_c = 2000$ K and $P_c = 0.2$ MPa, and the outflow temperature of the propellant from the tank is $T_i = 300$ K. The nozzle is designed by Equation (3.13) under the working condition of the thrust force $F_1 = 0.1$ N, and the calculated cross-sectional area of the nozzle throat is $A_t = 2.545 \times 10^{-7}$ m^2.

When the solar energy value is kept constant, the thrust is changed only by changing the propellant flow rate, and the cross-sectional area of the nozzle throat is kept constant. The calculation results are shown in Table 3.3.

An analysis of the calculation results shows that when the solar energy remains constant, it is not advisable to adjust the thrust by changing the propellant flow rate. When the thrust force increases from 0.1 to 0.5 N, the propellant mass flow rate gradually increases, the total thrust chamber temperature decreases sharply from 2000 to 760 K and then to 360 K, and the total thrust chamber pressure Pc increases sharply from 0.2 to 1.0 MPa.

3.6.2   Changing the Solar Energy While Maintaining the Propellant Flow Rate

When hydrogen is selected as a working fluid, the thruster can obtain greater thrust at high chamber pressure; therefore, when the thrust $F = 0.5$ N, the designed thrust chamber pressure $P_c = 0.8$ MPa, and the thrust chamber temperature is $T_c = 2000$ K. The calculated cross-sectional area of the nozzle throat is $A_t = 3.182 \times 10^{-7}$ m^2, and the propellant flow rate is $\dot{m} = 5.890 \times 10^{-5}$ kg/s. When the propellant flow rate is constant, the thrust force is changed only by adjusting the solar energy, and the results are shown in Table 3.4.

Table 3.4 Calculation results under each working condition with a constant flow rate

Working condition	0.1 N	0.16 N	0.2 N	0.3 N	0.4 N	0.5 N
Mass flow rate ($\times 10^{-5}$ kg/s)	3.65	5.89	5.89	5.89	5.89	5.89
Solar energy (W)	0	0	149	679	1422	2400
Exhaust speed (m/s)	2736	2736	3395	5093	6791	8488
Total temperature (K)	300	300	320	720	1280	2000
Overall pressure (MPa)	0.20	0.20	0.32	0.48	0.64	0.8

Table 3.5 Constant nozzle inlet temperature

Working condition	0.1 N	0.2 N	0.3 N	0.4 N	0.5 N
Mass flow rate ($\times 10^{-5}$ kg/s)	1.178	2.356	3.534	4.713	5.891
Solar energy (W)	475.4	950.7	1426	1901	2376
Exhaust speed (m/s)	8488	8488	8488	8488	8488
Total temperature (K)	2000	2000	2000	2000	2000
Overall pressure (MPa)	0.1	0.2	0.3	0.4	0.5

When the flow rate is kept constant and a cold gas propulsion system is used, the thrust force cannot reach the designed value of 0.1 N. At this moment, even if there is no incident sunlight, the minimum thrust force is 0.16 N. When comparing the adjustment of solar energy with that of the mass flow rate, the variation in the total temperature of the thrust chamber is relatively smooth, there is no abrupt change, and the variation in the total pressure is uniform.

3.6.3 Maintaining a Constant Nozzle Inlet Temperature (2000 K)

To ensure the normal operation of the thruster, the working fluid is heated to the normal working temperature (>2000 K). The calculation results are shown in Table 3.5. The designed cross-sectional area of the nozzle throat is $A_t = 5.07 \times 10^{-7}$ m^2. When the thrust force increases, the flow rate and solar energy gradually increase, and changing the flow rate and solar energy simultaneously is feasible.

Chapter 4
Radiation Heat Transfer of Absorption Cavity and Secondary Concentrator

4.1 Introduction

In this chapter, the design is based on the foreign refractive secondary concentrator (RSC) structures, the specific dimensions of the concentrator are modified according to the actual situation, an integrated simulation of the radiation and flow heat transfer in the RSC and the absorption cavity is performed, and the thermal stress of the RSC is analyzed.

4.2 Physical Models and Boundary Conditions

4.2.1 Physical Model

The RSC is a nonimaging concentrating system that concentrates the incident light into the absorber through the reflection and total internal reflection (TIF) between different media. The RSC is composed of a lens and an energy extractor. The lens has an axisymmetric structure, while the energy extractor has a triangular pyramid structure. After the sunlight enters the concentrator after being reflected by the lens, it is transmitted according to the principle of TIF. Since the concentrated light is always transmitted inside the concentrator, the energy output loss is relatively small. Finally, the energy reflected by the energy extractor is transmitted out of the concentrator to heat the wall of the absorption cavity. The integrated design of the regenerative cooled RSC and the thrust chamber and the propellant flow path are shown in Fig. 4.1. After entering the thruster, the propellant first flows through a porous sleeve to achieve even flow splitting so that the gas flow and temperature distribution around the RSC are uniform. The propellant enters the absorption cavity after passing through the sleeve, the RSC in the absorption cavity bears a relatively higher temperature load, and the regenerative cooling in this area reduces the RSC temperature through convective heat

© National University of Defense Technology Press 2025
M. Huang et al., *Solar Thermal Thruster*, https://doi.org/10.1007/978-981-97-7490-6_4

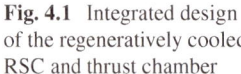
Fig. 4.1 Integrated design
of the regeneratively cooled
RSC and thrust chamber

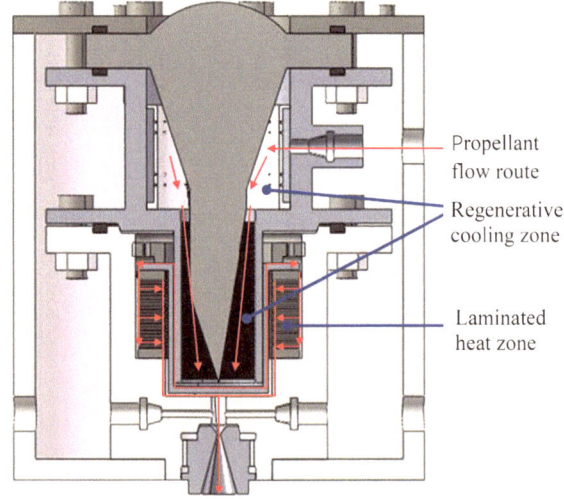

Propellant
flow route

Regenerative
cooling zone

Laminated
heat zone

transfer. Finally, the propellant enters the efficient laminated micro-heat exchange
runner through the opening at the bottom of the absorption cavity to be heated and
is finally discharged through the nozzle to generate thrust. The cooling channel is
the cavity between the absorption cavity and the RSC. Since the energy extractor
at the back end of the concentrator has a triangular pyramid structure, the cooling
channel is not a strict axisymmetric structure. This integrated design of regenerative
cooling not only has a significant cooling effect on the RSC but also increases the
temperature of the propellant entering the absorption cavity, which improves the
utilization efficiency of solar energy by the thruster.

Based on the physical model, the three media in the radiative heat transfer model
of the absorption cavity are shown in Fig. 4.2, with the RSC, the working fluid
hydrogen, and the Nb521 niobium-tungsten alloy as the absorption cavity material
from the inside to the outside. For the gaseous working fluid section, the upper part is
the working fluid inlet, and the bottom is the outlet, which conforms to the structural
design of the absorption cavity in Chapter 2.

4.2.2   Boundary Conditions

4.2.2.1   Interface Radiation Characteristics

The wall surface of the absorption cavity is an opaque interface and a diffuse surface
(diffuse emission and reflection), the interface temperature is T_1, the spectral emis-
sivity inside the interface is ε_λ, the spectral diffuse reflectance is ρ_λ^d, and the superscript d represents diffuse reflection. Let the subscript i of the angle represent inci-
dence, and the spectral reflected radiation force inside the absorption cavity interface

Fig. 4.2 Three media for the radiative heat transfer model of the absorption cavity

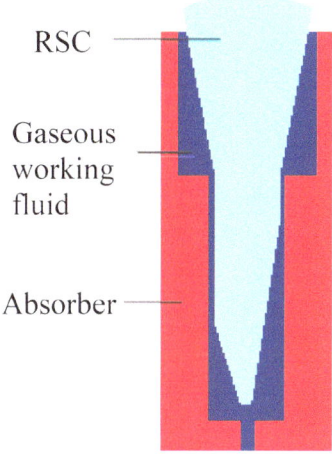

RSC

Gaseous working fluid

Absorber

is

$$
\rho_\lambda^d \int\limits_{2\pi} I_\lambda^-(0,\ \theta_i) \cos \theta_i d\Omega_i
$$

$$
= \rho_\lambda^d \int\limits_{\theta_i=\pi}^{\pi/2} \int\limits_{\Psi_i=0}^{2\pi} I_\lambda^-(0,\theta_i) \cos \theta_i \sin \theta_i d\theta_i d\Psi_i
$$

$$
= -\rho_\lambda^d 2\pi \int\limits_{\mu_i=-1}^{1} I_\lambda^-(0,\ -\mu_i)\mu_i d\mu_i
$$

$$
= 2\pi\rho_\lambda^d \int\limits_{0}^{1} I_\lambda^-(0,-\mu_i)\mu_i d\mu_i \tag{4.1}
$$

The effective radiation at the interface is the sum of the self-radiation and reflected radiation; then,

$$
I_\lambda^+(0) = \frac{1}{\pi} n_m^2 \varepsilon_\lambda \sigma T_1^4 + 2\rho_\lambda^d \int\limits_{0}^{1} I_\lambda^-(0,\ -\mu_i)\mu_i d\mu_i \tag{4.2}
$$

The RSC wall surfaces are all semitransparent interfaces, which are spectral selective surfaces; the interface is transparent for bands with wavelengths less than 5 μm and translucent for bands with wavelengths greater than 5 μm [115]. The interface radiation is the sum of the penetrating part of the projected radiation for the environment and the reflected radiation at the interface on the media side.

For diffuse surfaces,

$$I^+(0) = \left(\frac{n_m}{n_o}\right)^2 (1 - \rho_0^d) I_0(0) + 2\rho^d \int_0^1 I^-(0, -\mu)\mu d\mu \qquad (4.3)$$

If the interface is a mirror surface, then

$$I^+(0, \mu) = \left(\frac{n_m}{n_o}\right)^2 (1 - \rho_0^s) I_0(0, \mu_i) + 2\rho^s I^-(0, -\mu) \qquad (4.4)$$

$$\mu_i = \cos\theta_i \qquad (4.5)$$

where θ_i is the incident angle of the projected radiation for the environment.

4.2.2.2   Radiation Boundary Condition

The radiation boundary condition is a third-type boundary condition. The heat transfer coefficient h, the ambient temperature T_0, and the radiation characteristics on the inside of the interface is ρ_λ. The boundary condition is applicable to the opaque, semitransparent, and transparent interfaces involved in the model.

4.2.2.3   Flow Boundary Condition

For the regenerative cooling channel in the absorption cavity, the flow cross-section has no sharp contraction or expansion, so the pressure drop changes little. The mass flow rate at the propellant inlet is set to 0.000175 kg/s, and the outlet pressure is 0.7 MPa. The inlet temperature is set to 300 K.

4.2.3 Computational Grid

Because the RSC energy absorber has a triangular pyramid structure, the radiation transfer in the absorption cavity has 3D distribution characteristics, and the axisymmetric model is not suitable for simulating the distribution characteristics. Therefore, 3D modeling is used, and an unstructured grid is generated. A computational grid of the radiation heat transfer of the absorber is shown in Fig. 4.3.

Fig. 4.3 Computational grid
of absorber radiation heat
transfer

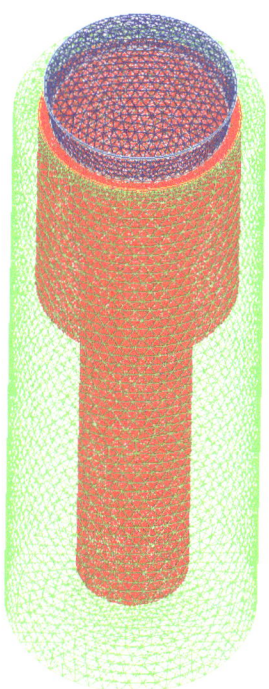

4.3   Radiation and Regenerative Cooling Processes
for the Absorption Cavity

4.3.1   Effect of Regenerative Cooling on the Absorption
Cavity

4.3.1.1   Design Without Regenerative Cooling

Regenerative cooling is an important energy utilization scheme proposed in this
design, and its main purpose is to prevent high temperatures and ruptures in the RSC
by absorbing too much heat during operation and to recover energy. In this section, a
simulation is performed on whether to use regenerative cooling technology to verify
the effectiveness of regenerative cooling.

First, calculations are performed for a case without regenerative cooling. There
is no gas flow in the absorption cavity. Figure 4.4 shows a temperature distribution
cloud map of the absorption cavity. The highest temperature of the absorber wall
exceeds 2400 K, and the high-temperature zone is in the lower half of the absorber,
which has reached the ideal working temperature of the absorber and can heat the
gaseous working fluid to a high temperature. The temperature distribution of the RSC
medium is shown in Fig. 4.5 after the extraction of the RSC medium. Figure 4.5 shows

that the temperature of the RSC medium is basically the same as the temperature of the absorption cavity, and the maximum temperature of the energy extractor at the bottom end reaches 2400 K, which already exceeds the operating temperatures that most media materials can withstand. For media such as single-crystal sapphire, the absorption of the solar spectrum smaller than 5 μm is very small and can be ignored, while the solar energy with wavelengths larger than 5 μm only accounts for approximately 0.5%. However, because the temperature of the heated absorption cavity is above 2400 K, the infrared radiation dominates; radiation with wavelengths longer than 5 μm accounts for more than 5%, which is 10 times higher compared to the solar spectrum, and all such radiation energy can be absorbed by the RSC. Therefore, the temperature of the RSC can continue to rise and eventually approach the wall temperature of the absorption cavity, which is higher than the withstanding temperature of the RSC. Therefore, the regenerative cooling structure of the adoption cavity is very important, which can take away the radiation energy absorbed by the RSC in time and prevent the RSC from cracking.

Under this design, the distributions of incident radiation and radiation temperature are shown in Figs. 4.6 and 4.7, respectively. The radiation heat transfer process is mainly concentrated at the bottom of the energy extractor, that is, the triangular pyramid. The radiation heat transfer at other locations is slight, which shows that the simulation results are reasonable. Due to the TIF of the upper part, the amount of solar radiation escaping is very small.

Fig. 4.4 Temperature distribution without regeneration cooling

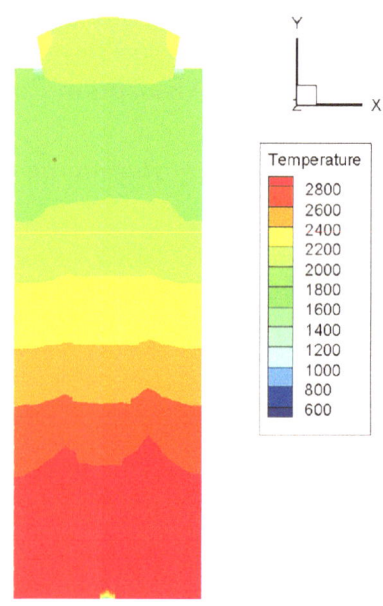

Fig. 4.5 RSC temperature distribution

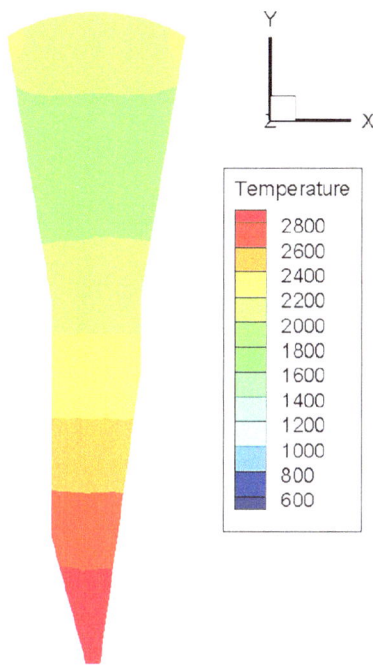

Fig. 4.6 Incident radiation distribution

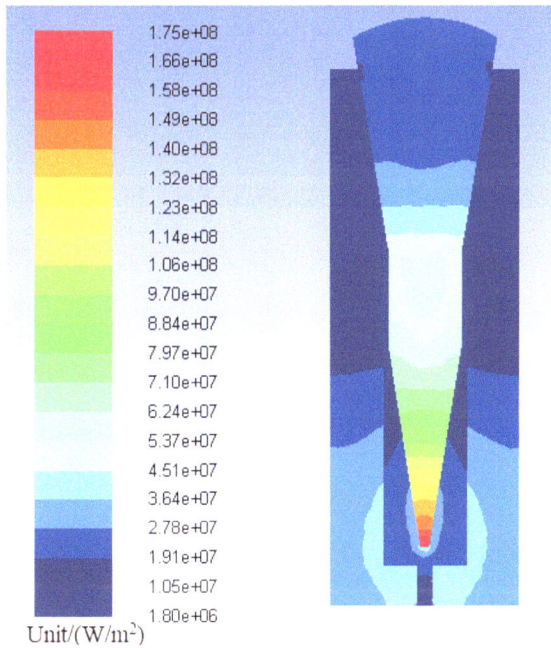

Fig. 4.7 Radiation
temperature distribution

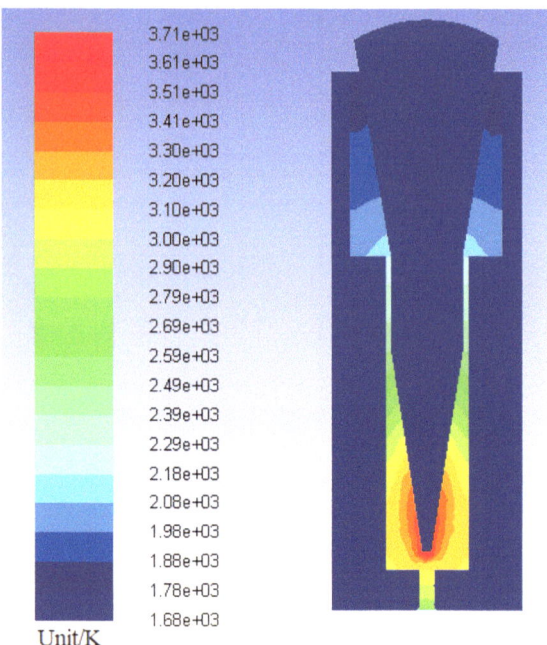

Unit/K
3.71e+03
3.61e+03
3.51e+03
3.41e+03
3.30e+03
3.20e+03
3.10e+03
3.00e+03
2.90e+03
2.79e+03
2.69e+03
2.59e+03
2.49e+03
2.39e+03
2.29e+03
2.18e+03
2.08e+03
1.98e+03
1.88e+03
1.78e+03
1.68e+03

4.3.1.2   Regenerative Cooling Design

For the case with regenerative cooling, Fig. 4.8 shows the overall temperature distribution characteristics of the RSC, the working fluid and the absorber, and Fig. 4.8a and
b show cross-sections of the thruster in different directions. Since the energy extractor
of the concentrator is approximately a triangular pyramid structure, the temperature
distribution characteristics of the cross-sections in different directions are slightly
different. The temperature distribution shows that because the energy extractor of
the RSC is located at the lower part of the absorption cavity, the temperature of
the absorber wall increases from top to bottom; the temperature of the hydrogen
working fluid is high when close to the absorber wall and decreases towards the
direction of the RSC. Therefore, the flow of working fluid should make full use of
the high-temperature zone at the bottom for heating. As shown in Fig. 4.8a and b,
the temperature of the propellant close to the RSC ranges between 800–1000 K,
which is significantly lower than that in the case without cooling measures, while
the temperature of the propellant near the absorber wall is higher at 1000–2200 K.
The regenerative cooling design also preheats the propellant. The temperature of the
propellant in the absorption cavity rises from 300 K at the inlet to 1000 K at the
outlet, thus improving the use efficiency of solar radiation energy. Without cooling
measures, the temperature of the concentrator can continue to rise, and correspondingly, the radiation loss to the outside world can also continue to increase. This issue
can be effectively alleviated by using the regenerative cooling design.

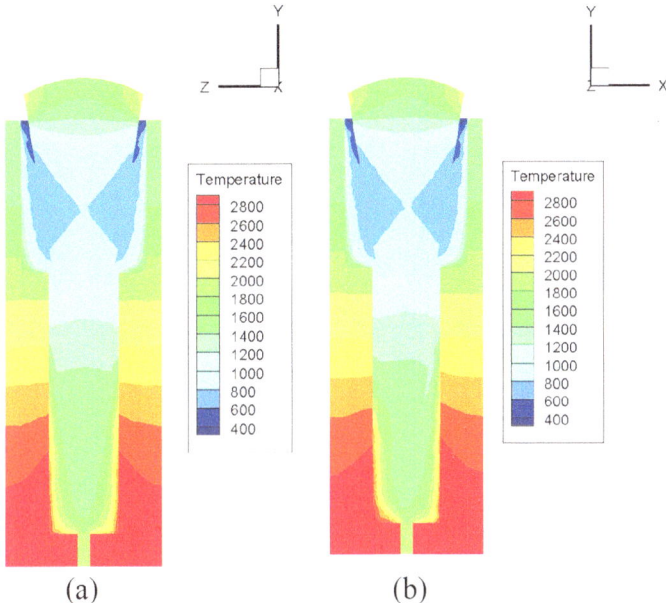

Fig. 4.8 Overall temperature distribution of RSC, working fluid, and absorber on cross-sections along different directions

The calculated temperature distribution of the absorption cavity wall and a comparison with the previous results show that the flow of the working fluid has little effect on the temperature of the absorption cavity, and the wall temperature of the absorber is still in the range of 2400 K to 2600 K, which can meet the temperatures needed for the propulsion system. This is because the heat transfer between the RSC and the wall of the absorption cavity is dominated by solar radiation, and the propellant in the absorption cavity is a transparent medium relative to the solar spectrum and has little effect on the transmission of concentrated solar radiation.

To facilitate the comparisons of the temperature distribution patterns of the RSC, working fluid hydrogen and absorber, Fig. 4.9a–c show the temperature distribution areas of the three media on the cross-section. The temperature distribution of the RSC gradually increases along the negative direction of the Y-axis, and the temperature of the energy extractor is the highest at 600–800 K. After regenerative cooling, the maximum temperature of the concentrator decreases from 2400 to 800 K. The heat for heating the RSC comes from the high-temperature wall surface of the absorption cavity rather than the direct absorption of solar energy. Because solar energy is also absorbed, the temperature of the upper part of the RSC is much lower than that of the energy extractor. The temperature of the working fluid is close to the temperature of the energy extractor. Therefore, most of the heat is transferred from the working fluid to the RSC through convective heat transfer. This design allows the RSC to operate for a long time without large thermal shocks. Therefore, general optical glass materials, especially quartz glass, can be used as alternative RSC materials.

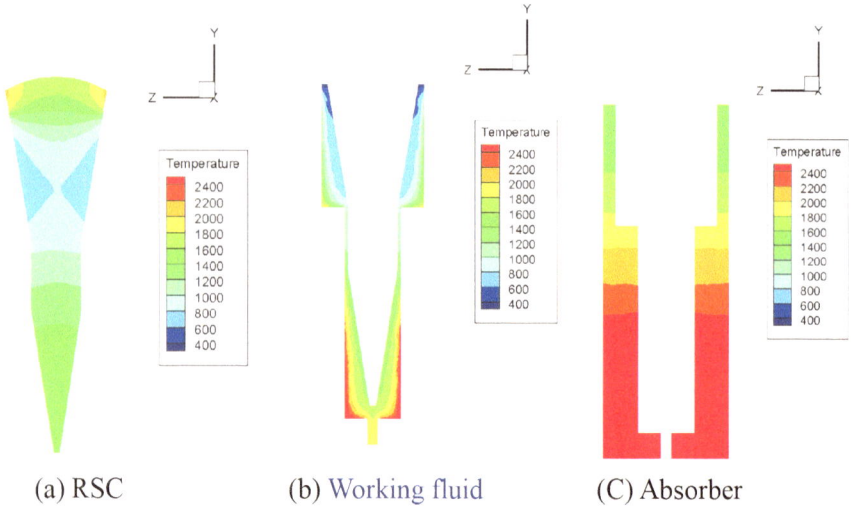

(a) RSC   (b) Working fluid   (C) Absorber

Fig. 4.9 Temperature distribution cloud map of RSC, working fluid and absorber

The temperature distributions for different axial positions perpendicular to the axial direction of the thruster are shown in Fig. 4.10. The average temperature of the sections gradually increases from the top to the bottom. The Y-axis coordinates at different positions are shown in Fig. 4.6. Because the energy extractor is a triangular pyramid, the cross-section for output thermal radiation presents an approximately equilateral triangle distribution pattern. Heat diffuses along the three directions normal to the triangular pyramid surface of the energy extractor, and the temperature rise also occurs along these three directions.

The radiation temperature and incident radiation distributions of the absorption cavity and RSC are shown in Figs. 4.11 and 4.12, respectively, with the maximum radiation temperature and incident radiation both appearing at the top of the energy extractor, which is in line with the light path transmission characteristics of the RSC. In the part before the energy extractor, due to the TIF of the light, no solar radiation is directly emitted from the upper interface, and all the solar radiation is concentrated at the energy extractor at the lower part of the RSC. A comparison with the design without regenerative cooling shows that the radiation temperature and the incident radiation distributions are close to each other, and the large difference in temperature distribution is completely caused by the regenerative cooling effect of the working fluid.

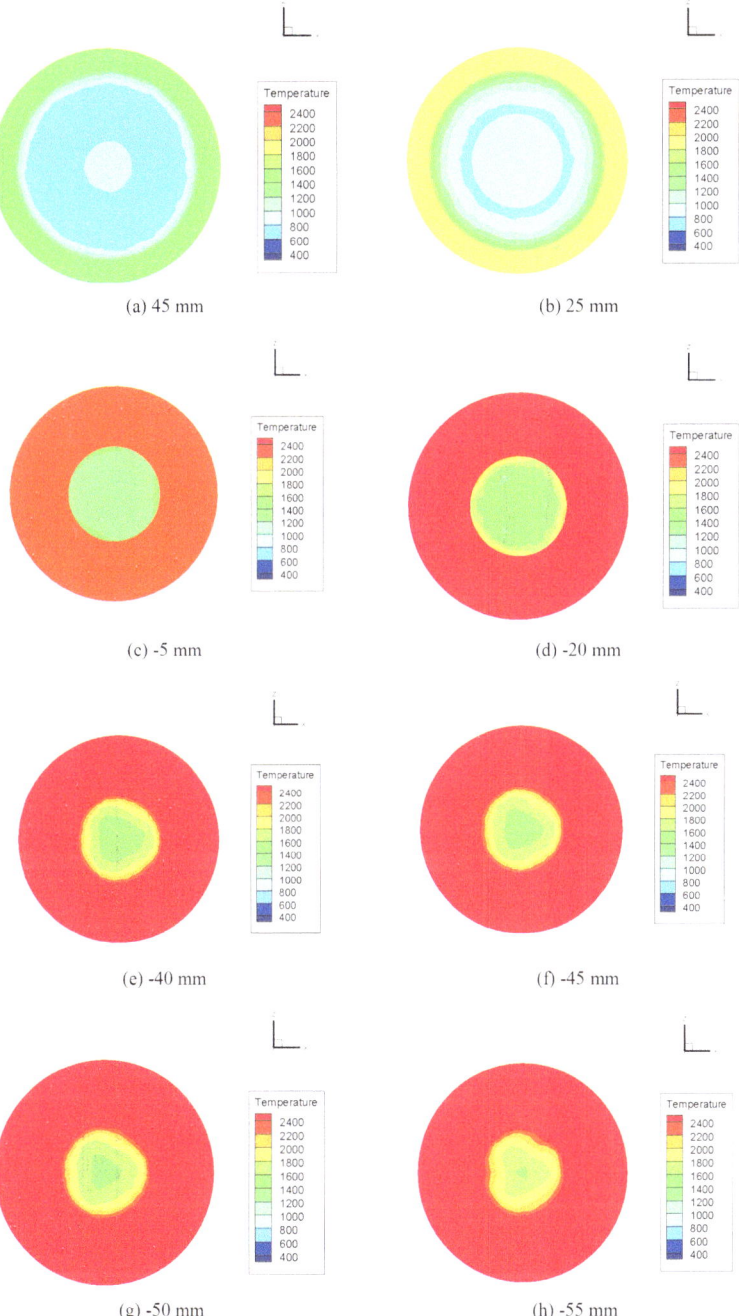

Fig. 4.10   Temperature distribution variations in sections along the axial direction in the thruster

Fig. 4.11 Radiation temperature distribution of the absorption cavity and RSC

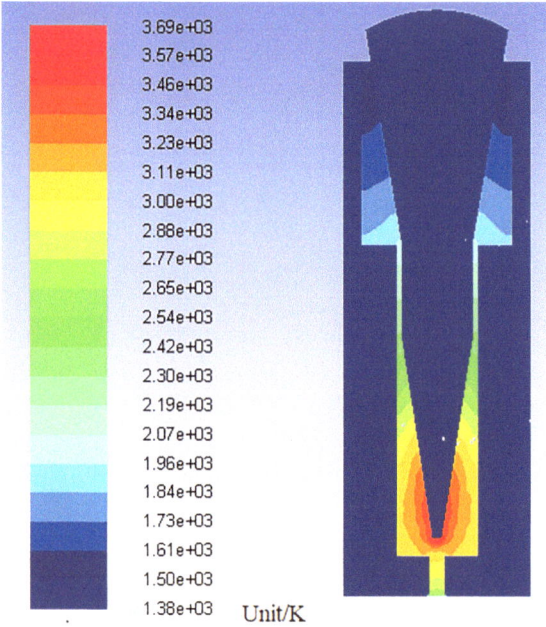

Fig. 4.12 Incident radiation distributions of absorption cavity and RSC

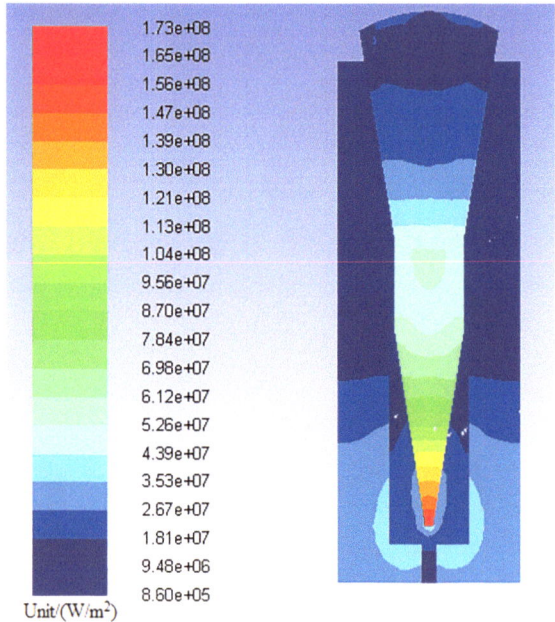

4.3.2   Effect of Absorption Coefficient

The absorption coefficient has a great impact on the RSC, which is a high-temperature core region. If the absorption coefficient of the RSC is large, the RSC can absorb a large amount of heat accordingly. In this case, the wall temperature of the absorption cavity will not be high. A larger absorption coefficient can affect the penetration of thermal radiation. Therefore, single-crystal materials with small absorption coefficients must be used as RSC materials. The thermophysical parameters of single-crystal materials that are easily obtained are shown in Table 4.1.

The absorption coefficient can be derived from data such as the spectral transmittance and emissivity. According to the relationship between the thermal radiation transmittance and the absorption coefficient,

$$UVT = \frac{I}{I_0} = e^{-\kappa_\lambda x} \tag{4.6}$$

where κ_λ is the absorption coefficient of the medium to spectrum λ and x is the thickness of the medium.

(1)   The absorption coefficient of 0.1 m^{-1}

The temperature distribution, incident radiation distribution and radiation temperature distribution when the absorption coefficient is 0.1 m^{-1} are shown in Figs. 4.13, 4.14 and 4.15, respectively. The temperature of the RSC is far lower than the temperature of the absorber wall. The temperature of the energy extractor is maintained at 600–800 K, which is lower than the melting point of general single–crystal materials and even lower than the melting point of quartz glass.

(2)   The absorption coefficient of 1 m^{-1}

The temperature distribution, incident radiation distribution and radiation temperature distribution with an absorption coefficient of 1 m^{-1} are shown in Figs. 4.16, 4.17 and 4.18, respectively. The temperature of the RSC is already high, with the highest temperature over 1600 K, and the wall temperature of the absorption cavity

Table 4.1   Thermophysical parameters of the main single-crystal materials

Material	Melting point (°C)	Reflective index	Thermal conductivity (W/mK)	Optical absorption cutoff wavelength (μm)
Al_2O_3 single crystal	2300	1.76	0.25 (300 K) 0.1 (1000 K) 0.06 (2300 K)	5
MgO single crystal	3000	1.76	0.6 (300 K) 0.08 (1500 K)	7
ZrO_2 single crystal	3000	2.16	0.1 (300 K)	6

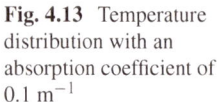

Fig. 4.13 Temperature distribution with an absorption coefficient of 0.1 m^{-1}

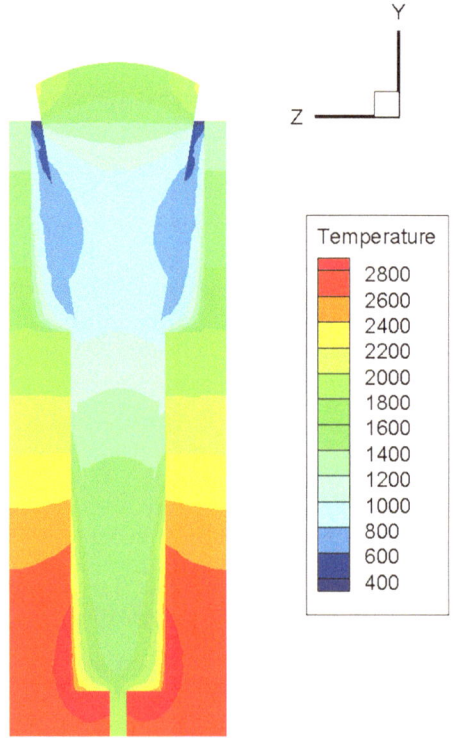

is higher than 2300 K; therefore, the material with an absorption coefficient of 1 m^{-1} can satisfy the working conditions under certain circumstances.

(3)   The absorption coefficient of 10 m^{-1}

The temperature distribution with an absorption coefficient of 10 m^{-1} is shown in Fig. 4.19. The temperature distribution shows that when the absorption coefficient of the medium reaches 10 m^{-1}, the absorption of solar radiation by the RSC is already very large; when the highest temperature of the absorption cavity wall is only 2200 K, the highest temperature inside the RSC already exceeds 2600 K. The incident radiation temperature distribution and radiation temperature distribution are shown in Figs. 4.20 and 4.21, respectively, and the maximum values of 4.5×10^7 W/m^2 and 2400 K appear at the top of the energy extractor, which are smaller than those of the first two working conditions, indicating that the RSC absorbs a large amount of sunlight, and the solar radiation transmitted through the RSC decreases.

Fig. 4.14 Incident radiation distribution with an absorption coefficient of 0.1 m^{-1}

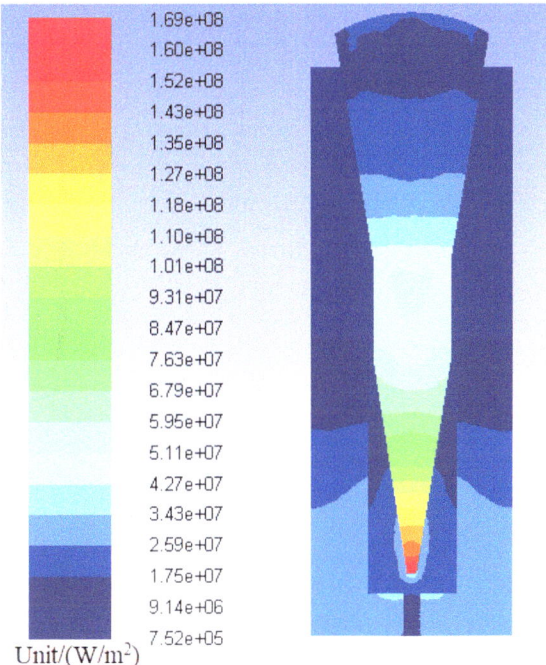

Fig. 4.15 Radiation temperature distribution with an absorption coefficient of 0.1 m^{-1}

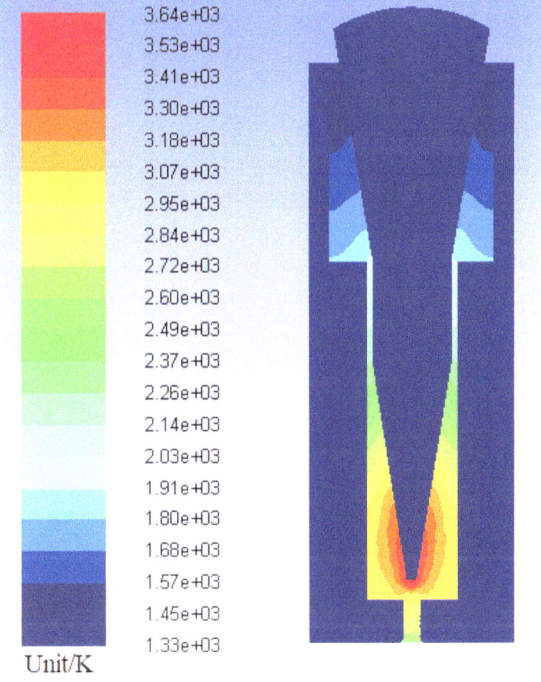

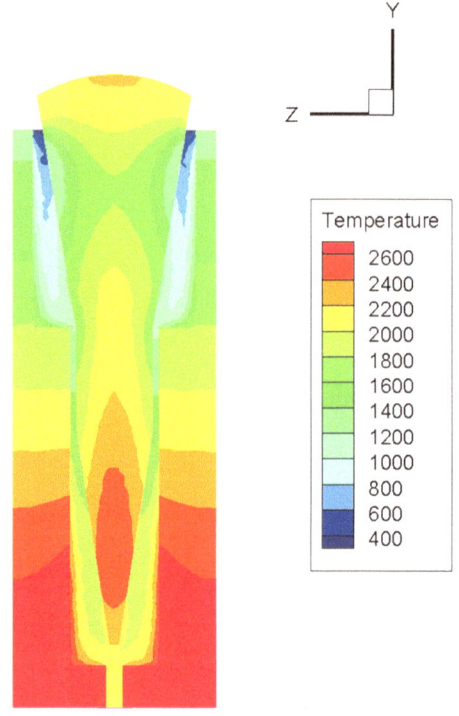

Fig. 4.16 Temperature distribution with an absorption coefficient of $1\ \mathrm{m}^{-1}$

4.3.3   Effect of Using a Non-gray Semitransparent Medium

Since the RSC medium has different absorption coefficients of the solar spectrum at different wavelengths, the spectral band approximation model (SBAM) can be used for analysis and processing [117]. Figure 4.22 shows the radiation energy distribution of the solar spectrum. Solar spectral energy is mainly concentrated in the visible and near-infrared regions, and the RSC exhibits very little absorption to this spectral band since it is the main transmission band. The blackbody radiation function is defined as

$$f(\lambda T) = \int\limits_{0}^{\lambda T} \frac{E_{b\lambda}}{\sigma T^5} \mathrm{d}(\lambda T) \tag{4.7}$$

According to the blackbody radiation function, the blackbody radiation at each wavelength can be calculated as a percentage of the blackbody radiation at the same temperature. The solar energy for spectral band larger than 5 μm accounts for less than 5% of the solar energy. This part of the energy can be absorbed by the single-crystal material, so more accurate temperature distribution characteristics can be obtained by using the segmented SBAM.

Fig. 4.17 Incident radiation distribution with an absorption coefficient of 1 m^{-1}

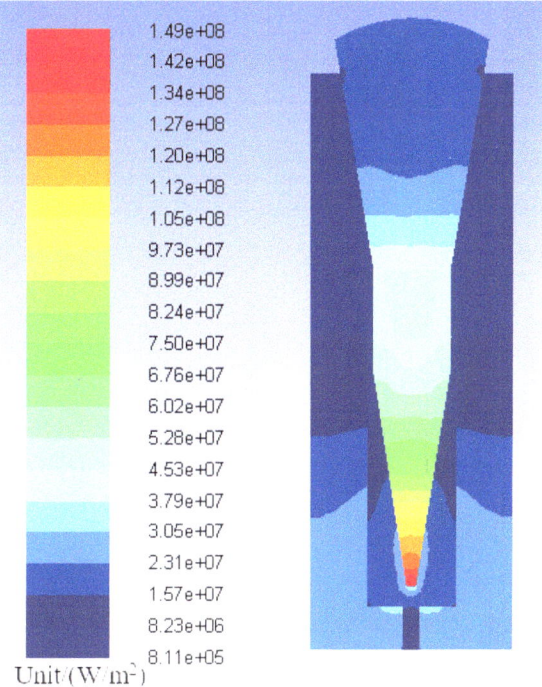

Fig. 4.18 Radiation temperature distribution with an absorption coefficient of 1 m^{-1}

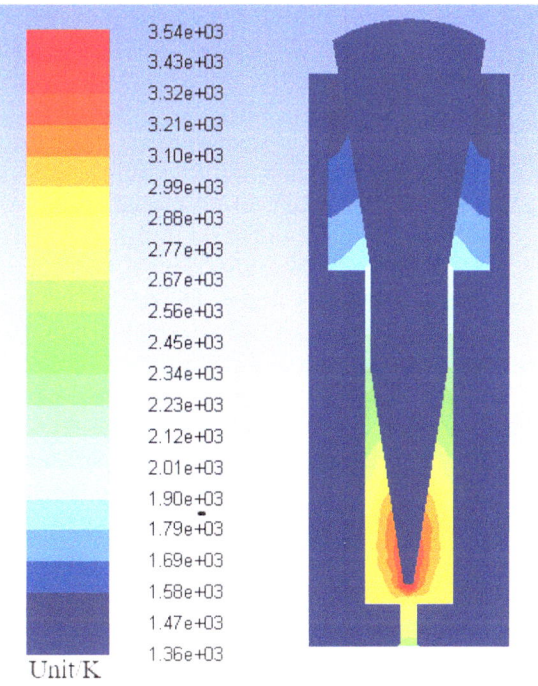

Fig. 4.19 Temperature distribution with an absorption coefficient of 10 m^{-1}

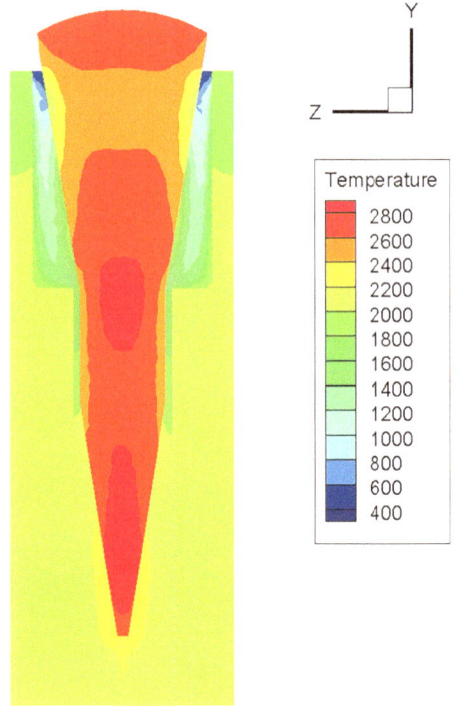

The overall trend of the solar spectral absorption of a single–crystal sapphire is shown in Fig. 4.23. The absorption coefficient increases with increasing wavelength, and the solar absorption for wavelengths greater than 5 μm is already strong. According to the spectral characteristics of the single-crystal sapphire, the variation function of the absorption coefficient versus temperature is established in different spectral bands for subsequent simulation.

Figure 4.24a shows the result of the gray semitransparent medium model, and Fig. 4.24b shows the result after using the non-gray semitransparent medium model. The RSC temperature calculated by the latter is low, and the highest temperature is concentrated in the absorption cavity, while the highest temperature calculated by the former is concentrated in the RSC. Because the spectral absorption characteristics are not considered, all the solar radiation is absorbed by the RSC medium according to the proportion corresponding to the absorption coefficient; therefore, the RSC temperature is the highest. With increasing temperature, the absorption coefficient of the medium further increases, resulting in a decrease in the projection characteristics, so the temperature of the absorption cavity is not as high as that of the RSC. After using the SBAM, the solar energy corresponding to each spectral band is calculated since the different absorption coefficients of the medium for different spectra are considered. Most of the alternative media are transparent to the visible and near-infrared light, and even if they are opaque, the absorption coefficient is also small, so the calculated results are more in line with the actual working conditions.

Fig. 4.20 Incident radiation distribution with an absorption coefficient of 10 m^{-1}

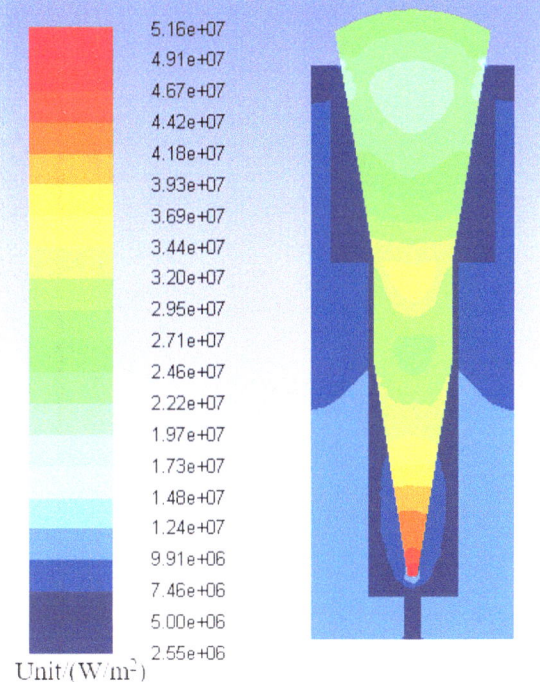

| 5.16e+07 |
| 4.91e+07 |
| 4.67e+07 |
| 4.42e+07 |
| 4.18e+07 |
| 3.93e+07 |
| 3.69e+07 |
| 3.44e+07 |
| 3.20e+07 |
| 2.95e+07 |
| 2.71e+07 |
| 2.46e+07 |
| 2.22e+07 |
| 1.97e+07 |
| 1.73e+07 |
| 1.48e+07 |
| 1.24e+07 |
| 9.91e+06 |
| 7.46e+06 |
| 5.00e+06 |
| 2.55e+06 |

Unit/(W/m^2)

Fig. 4.21 Radiation temperature distribution with an absorption coefficient of 10 m^{-1}

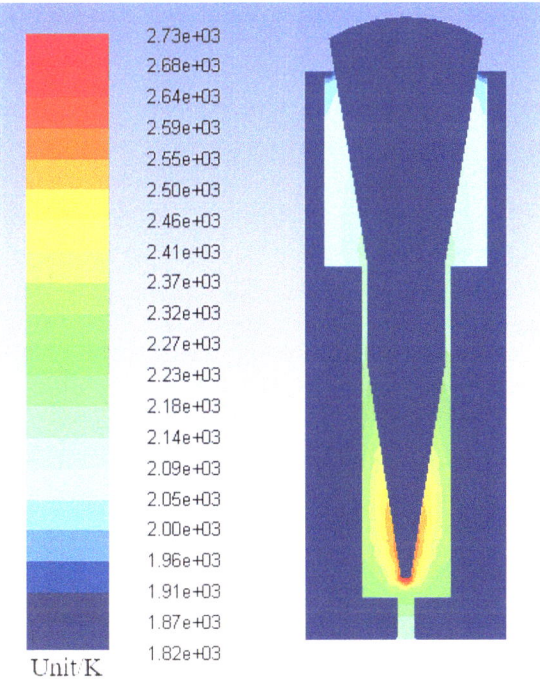

| 2.73e+03 |
| 2.68e+03 |
| 2.64e+03 |
| 2.59e+03 |
| 2.55e+03 |
| 2.50e+03 |
| 2.46e+03 |
| 2.41e+03 |
| 2.37e+03 |
| 2.32e+03 |
| 2.27e+03 |
| 2.23e+03 |
| 2.18e+03 |
| 2.14e+03 |
| 2.09e+03 |
| 2.05e+03 |
| 2.00e+03 |
| 1.96e+03 |
| 1.91e+03 |
| 1.87e+03 |
| 1.82e+03 |

Unit/K

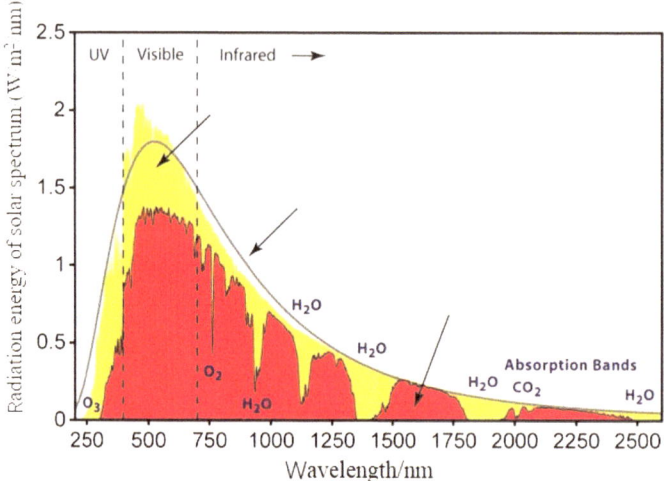

Fig. 4.22 Radiation energy distribution of solar spectrum

Fig. 4.23 Overall trend of
solar absorption of
single-crystal sapphire

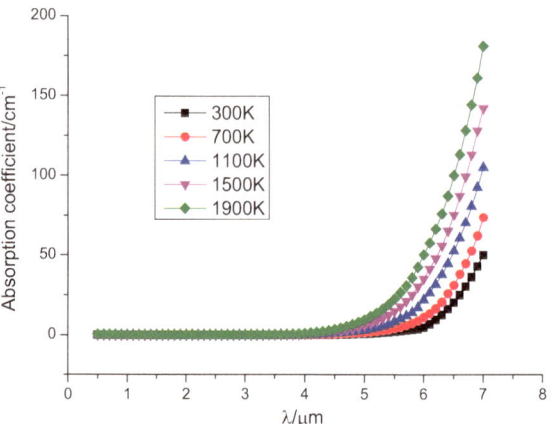

The temperature distribution cloud maps show that, even in Fig. 4.24b with lower temperatures, the highest temperature of the RSC is in the range of 1600–1800 K, which is still too high. Therefore, for the RSC medium, a material with a low spectral absorption coefficient should be selected. Single-crystal materials can satisfy this condition, but the cost is too high.

A comparison of the absorption coefficients of quartz glass and single-crystal sapphire materials is shown in Fig. 4.25, and the absorption coefficients of the two materials in the near-infrared spectrum are compared. At the same wavelength, the absorption coefficient of quartz glass is about 100 times higher than that of single-crystal sapphire material. The excellent optical properties of single-crystal materials are unmatched by other materials.

Fig. 4.24 Comparison of the temperature distribution between the gray semitransparent medium model and the non-gray semitransparent medium model

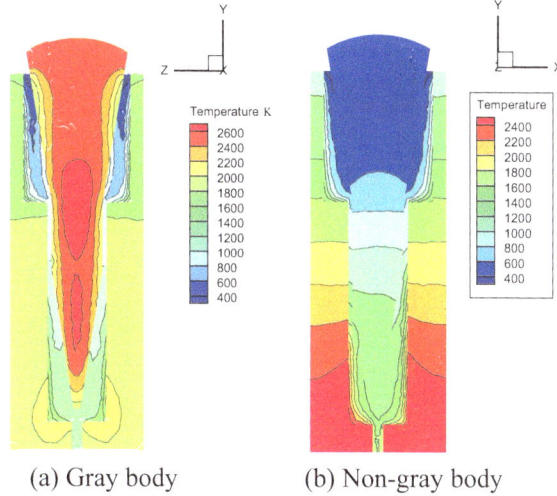

(a) Gray body   (b) Non-gray body

Fig. 4.25 Comparison of the absorption coefficients of quartz glass and single-crystal sapphire material

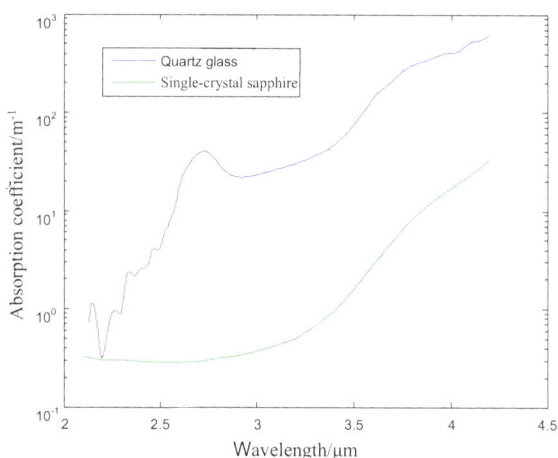

4.4   Thermal Stress Simulation of RSC

After loading the simulation results of the flow field on the RSC, the temperature and stress distribution characteristics of the RSC can be obtained. To study the cause of RSC cracking, the instantaneous temperature distribution of the RSC is calculated, and three typical nodes are selected for analysis, as shown in Fig. 4.26. When there is no regenerative cooling structure, the temperature distribution cloud map and the temperature variations at typical nodes during the RSC working process are shown in Fig. 4.27. The temperature cloud map shows that the temperature difference at the two ends of the RSC is large, so the thermal stress at the connection between the concentrator and energy extractor in the middle is the largest. The temperature of node

1 reaches equilibrium quickly, while the temperatures of nodes 2 and 3 gradually rise. The highest temperature of the RSC exceeds 2200 K, while the critical temperature of single-crystal sapphire is 2300 K, so cooling measures must be taken.

Figure 4.28 shows the temperature distribution and the temperature variations at typical nodes on the RSC with regenerative cooling. Figure 4.28 shows that the maximum temperature of the RSC decreases to 1600 K. The temperature variation patterns at the three nodes are the same as those in the previous analysis, with the maximum temperature decreasing by approximately 600 K. Figure 4.28a shows the thermal stress distribution inside the RSC. Figure 4.28a shows that the thermal stress

Fig. 4.26 Three typical nodes selected on RSC

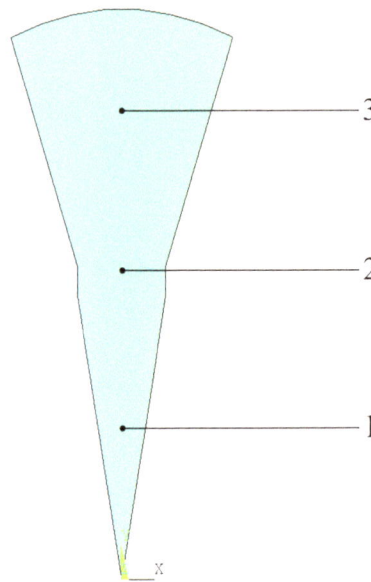

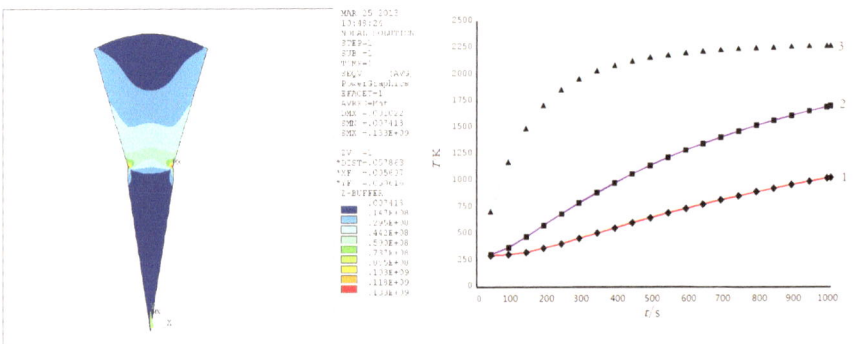

(a) Thermal stress distribution cloud map   (b) Temperature variation curve

Fig. 4.27 Internal thermal stress distribution of the RSC without regenerative cooling

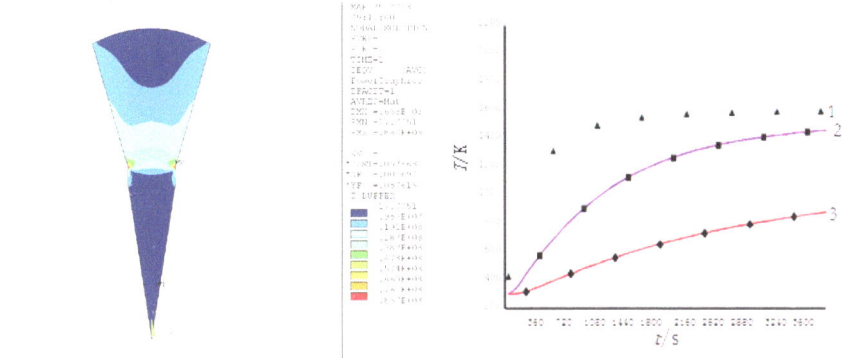

(a) Thermal stress distribution cloud map   (b) Temperature variation curve

Fig. 4.28 Internal thermal stress distribution of the RSC with regenerative cooling

is concentrated at the neck of the RSC, with a maximum stress of 113 MPa without regenerative cooling, and the thermal stress in most areas of the neck is 44–59 MPa. When regenerative cooling is used, the maximum stress is 86 MPa, and the thermal stress in most areas of the neck is 28–38 MPa. A comparison with the literature [48] shows that the thermal stress after regenerative cooling is lower than the measured thermal stress when the RSC is ruptured, at 44–65 MPa, which shows that the use of regenerative cooling design can improve the stability and reliability of the RSC.

Chapter 5
Simulation and Optimization Design of a Laminated Heat Exchanger Core for a Solar Thermal Thruster

5.1 Introduction

In this chapter, a laminated structure is used to design a high-efficiency heat exchanger core, and the optimal design scheme is obtained through a simulation and comparison analysis of the designed control runner length and runner cross-sectional area.

5.2 Temperature Distribution Characteristics after Heating Using Laminated Structures

5.2.1 Physical Models and Calculation Method

In the solar thermal thruster scheme, an improvement is made based on the laminated structure, i.e., a laminated microchannel structure and flow shunting are used to increase the heat exchange area between the propellant and the thrust chamber wall, thus improving the convective heat transfer in the heat exchange channel. As a result, the propellant is fully heated in the thrust chamber. The overall structure is shown in Fig. 5.1.

The design thickness of a single laminate is 2 mm, and the diameter of the control runner is 0.16 mm. The heat transfer area of the propellant between the laminates is 5–10 times larger than that of a spiral runner under the same conditions. The propellant flows along the outer edge of the laminate, passes through the control runner, and finally flows into the nozzle through the semicircular channel on the inner edge. The inner edge of the laminate is tightly bonded to the high-temperature wall of the absorption cavity, as shown in Fig. 5.2.

© National University of Defense Technology Press 2025
M. Huang et al., *Solar Thermal Thruster*, https://doi.org/10.1007/978-981-97-7490-6_5

Fig. 5.1 Overall structure
diagram of the laminated
enhanced heat exchange
channel

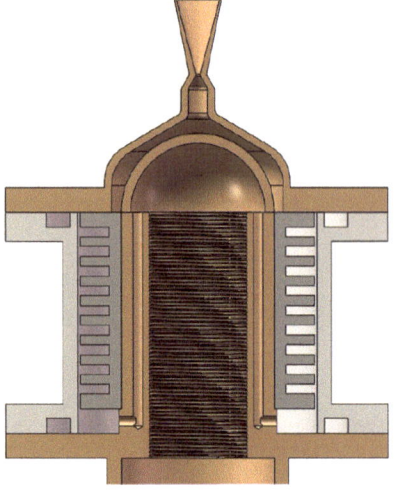

Fig. 5.2 Single laminate
microchannel structure

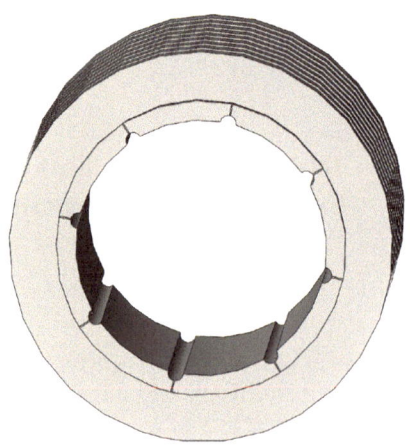

After considering the overall mass and heat exchange efficiency of the thruster, the designed structure of the heat exchanger core is shown in Fig. 5.3. Several laminates are stacked together, and the working fluid is further heated through the heat exchange microchannels in the middle. The propellant/working fluid inflows along the outside of the heat exchanger core and outflows at the inside. The inner wall of the heat exchanger core is seamlessly coupled to the high-temperature outer wall of the absorption cavity, and the energy is transferred from the high-temperature absorption cavity wall to the heat exchanger core laminates by heat conduction.

The number of laminates is designed to be 9, and the inner and outer diameters of the laminate are 32 mm and 49 mm, respectively. The radial length of the laminate is 40 mm, the length of the control runner is 2.5 mm, and there are eight control

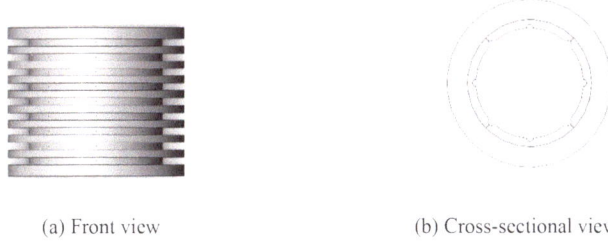

(a) Front view   (b) Cross-sectional view

Fig. 5.3 Heat exchanger core model

runners evenly distributed in the circumferential direction in each laminate. The overall structure is shown in Fig. 5.4.

Due to the axisymmetric characteristics of the laminate structure, to reduce the calculation volume, half of the single runners in the laminate structure are taken in the calculation. The upper, lower, left, and right sides of the model are symmetric planes, and the circumferential angle in the model is 22.5°. In reality, the control

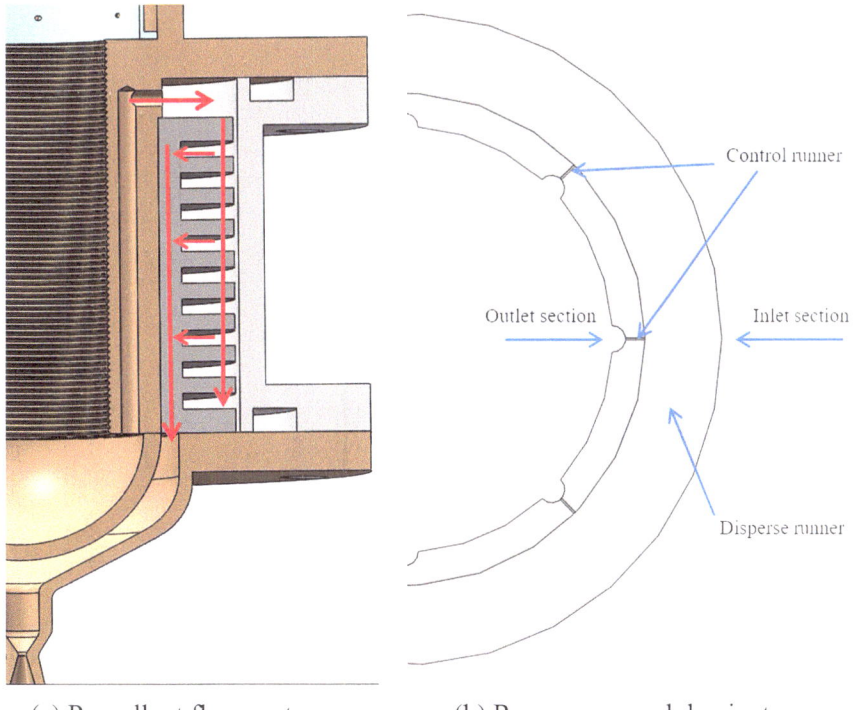

(a) Propellant flow route   (b) Runners on each laminate

Fig. 5.4 Overall structure of the laminated heat exchange channel

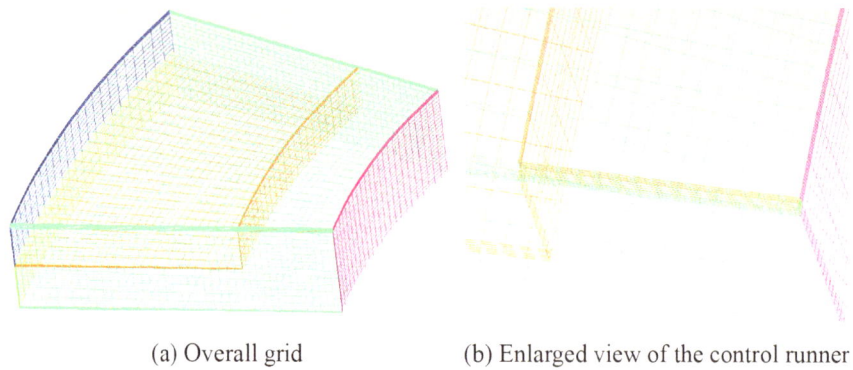

(a) Overall grid   (b) Enlarged view of the control runner

Fig. 5.5 Grid for the fluid–solid coupling model

runner is a circular channel. Because of the symmetry of the structure, only a quarter of the runner is selected for the simulation. The size of the runners is relatively small; therefore, square runners with the same cross-sectional area are used instead of circular runners in the model, which is convenient for generating meshes for calculation and simulation. The origin of the model is the physical center of the laminate. The centerline of the runner is set as the x-axis, which extends outward along the radial direction of the laminate. The y-axis is perpendicular to the z-axis in the plane of the runner. The z-axis is directed from the symmetry plane of the runner to the symmetry plane of the laminate. A structural grid is used to delineate the model, and the control runners are encrypted, with a total of 47,060 elements. A 3D grid model is used for the fluid–solid coupling, as shown in Fig. 5.5.

The physical properties and boundary conditions of the laminates and propellant of the above-described fluid-solid coupling system are applied to the discrete model; the steady state is obtained after the iterative calculations, and the temperature field and flow field of the laminated heat exchanger core of the solar thermal thruster are obtained.

5.2.2 Single-Channel Distribution Characteristics

After the fluid-solid coupling calculation, the overall 3D temperature distribution of the single-channel model is shown in Fig. 5.6. The temperature variations of the fluid part and the solid part are significantly different.

A cross-sectional analysis of the fluid temperature distribution in the laminate channel is performed along the direction perpendicular to the z-axis. Figure 5.7a–c shows the temperature distribution variation pattern from the channel center to the laminate wall, and the temperature of the working fluid in the channel shows a gradual increasing trend from the center to the wall. The heat exchange between the laminates and the propellant is mainly performed in the dispersed channel, and the

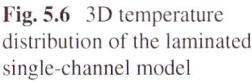

Fig. 5.6 3D temperature distribution of the laminated single-channel model

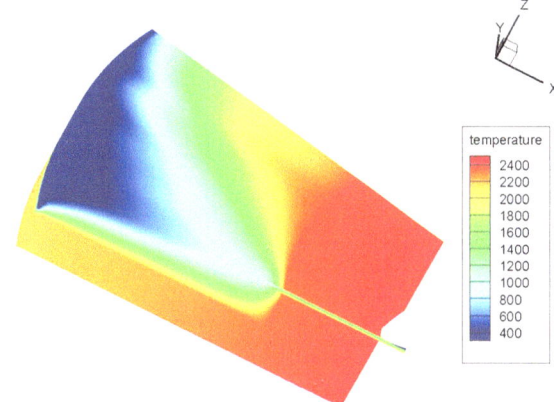

temperature variation of the working fluid in the control runner is relatively small. Figure 5.7a shows that the mainstream temperature of the runner center is heated to 1800 K at the laminate exit, and the temperature of the runner edge is above 2200 K, which is close to the boundary condition of the laminate wall temperature. Figure 5.7b shows the temperature distribution at 0.03 mm, and the changes are more pronounced compared to the symmetric plane. Figure 5.7c shows the temperature distribution of the control runner near the laminate wall. The propellant temperature at the outlet of the runner exceeds 2300 K, with the highest temperature being 2393 K, which is close to the wall temperature of 2400 K. After flowing through the laminates, the propellant enters the vertical interlayer and is further heated, thus raising the overall temperature of the propellant to exceed 2300 K. Simulation results from the relevant literature show that when a helical channel structure is used and the wall temperature is 2300 K, the highest temperature at the end of the helical channel and the interlayer section exceeds 2100 K, but does not reach 2200 K, indicating the superior thermal efficiency of the laminate heat transfer structure.

Figure 5.8 shows the temperature distribution profile of the solid part of the laminate along the direction perpendicular to y. Figure 5.8 shows that the laminate temperature has a gradient change from 2000 to 2400 K. In previous simulations, the temperature of the thruster wall is directly given as a constant temperature boundary condition; therefore, there is a certain error. The fluid temperature and the solid temperature largely interact with each other by many factors. The temperature of the cold end of the solid is 2000 K without considering the heat exchange between the system and the outside world. The heat exchange between the working fluid in the laminate dispersed runner and the laminate wall is sufficient, and the heating effect of the laminates is superior.

A cross-sectional analysis of the fluid speed distribution in the laminate runner is performed along the direction perpendicular to the z-axis. Figure 5.9a–c shows the speed distribution pattern from the runner center to the laminate wall. Due to the boundary layer, the speed of the working fluid in the runner generally shows a gradual decreasing trend from the center to the wall. The control runner of the

Fig. 5.7 Temperature
distribution of the fluid
cross-section in the laminate
channel

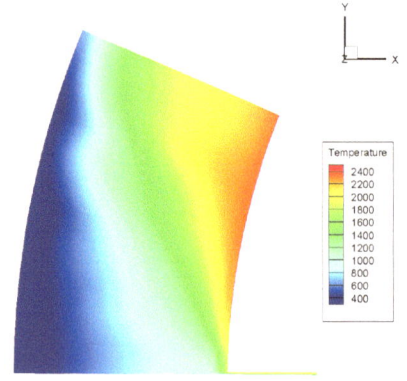

(a) Z-axis coordinate: 0, runner symmetric plane

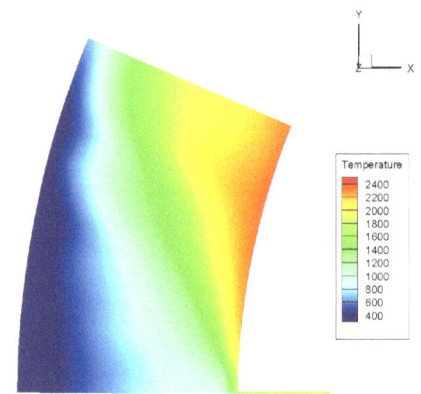

(b) Z-axis coordinate: 0.03 mm

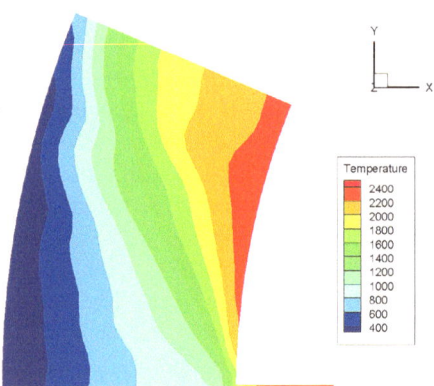

(c) Z-axis coordinate: 0.08 mm

Fig. 5.8 Temperature
distribution of the solid part
of the laminate

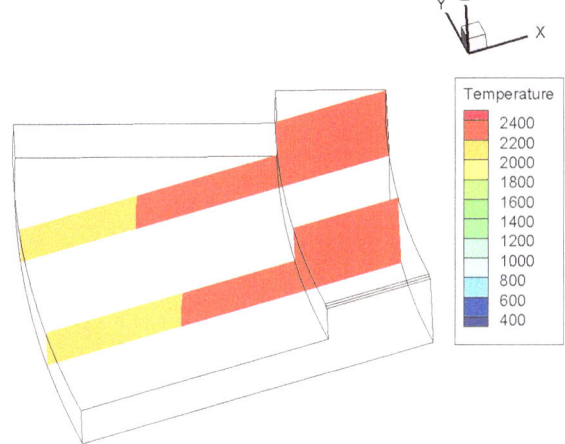

laminates accelerates the flow of propellant. The maximum speed at the center of
the control runner reaches 2500 m/s, and the local Mach number varies between 0.6
and 0.7.

The pressure distribution in the laminate runner is shown in Fig. 5.10.

5.3   Effect of Laminated Structure Parameters on the Heating Effect of the Heat Exchanger Core

5.3.1   Effect of Control Runner Length

The structural design of the laminates greatly affects the heating effect of the heat
exchanger core. The simulation study compares and analyses the effects of the control
runner length and cross-sectional area and the length ratio of the dispersion region
to the control runner. First, laminate models with different control runner lengths
are created to analyze the effect of the control runner length on the heating effect of
the laminates. At the same time, simulation is performed on the laminate structure
without control runners, which is equivalent to the condition when the control runner
length is zero, and the simulation results are compared and analyzed. To obtain the
optimal size of the laminate structure, the comparison focuses on the effect of the
control runner length on parameters such as heating temperature and speed.

5.3.1.1   Distribution Characteristics of Fluid Area

Figure 5.11 shows the distribution of the propellant temperature on the symmetric
plane of the laminate under different control runner lengths. A comparison of the

Fig. 5.9 Speed distribution of the propellant in the laminated runner sections

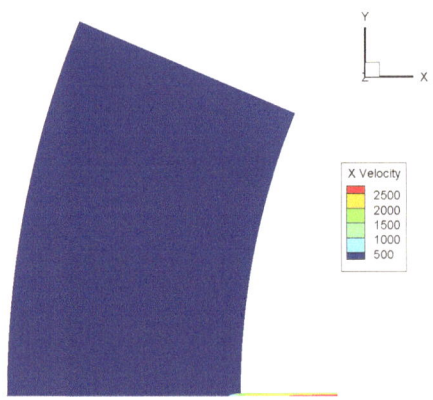

(a) Z-axis coordinate: 0, runner symmetric plane

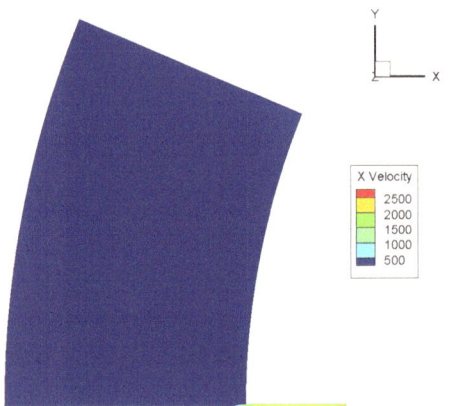

(b) Z-axis coordinate: 0.03 mm

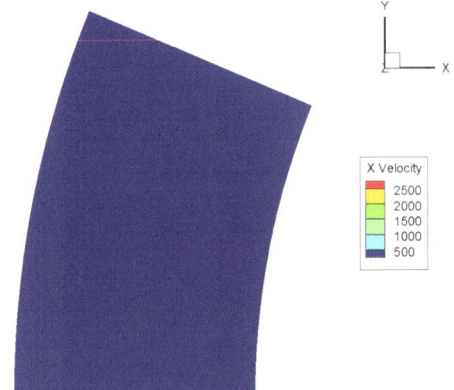

(c) Z-axis coordinate: 0.05 mm

Fig. 5.10 Pressure distribution of the fluid cross-section in the laminate runner

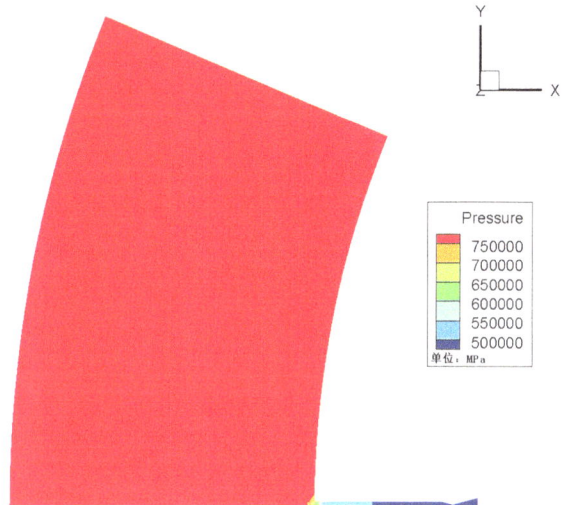

temperature distribution shows that the heating effect of the laminate on propellant without the control runner is the worst, with the highest temperature only being 1500 K after heating. After using control runners, the temperature of the propellant after heating significantly increases. In the 5 mm long control runner, as shown in Fig. 5.11f, the propellant is compressed and heated to a higher temperature, approximately 1800 K, before entering the control runner. After passing through the control runner, the propellant continues to be heated above 2200 K in the dispersion region behind the control runner.

The calculation results of the laminate structure with different control runner lengths are compared and analyzed, as shown in Table 5.1. The average temperature and the average total temperature of the cross-section of the heat exchanger core outlet are calculated by area integration. The overall temperature and total temperature of the heat exchanger core outlet are higher when the control runner length is 2–3 mm.

Figure 5.12 shows the trend of the average temperature and average total temperature at the outlet of the heat exchanger core under different control runner lengths. The 2 mm design scheme has the optimal average temperatures among several designs.

Figure 5.13 shows the temperature distribution at the cross-section of the heat exchanger core outlet under different control runner lengths, and the trend of the temperature distribution of several designs can be clearly observed in the figure. With the increase in the control runner length, the proportion of the high-temperature area of the cross-section at the outlet gradually increases, since the reduction in the dispersion area length can reduce the expansion effect of the propellant after passing through the control runner. The propellant temperature continues to decrease at the outlet of the 5 mm runner, indicating that the continued lengthening of the control runner reduces the heating effect on the propellant. A comparison of the temperature distribution of the cross-section at the heat exchanger core outlet under different control runner lengths is shown in Fig. 5.14. Figure 5.14 shows that although the

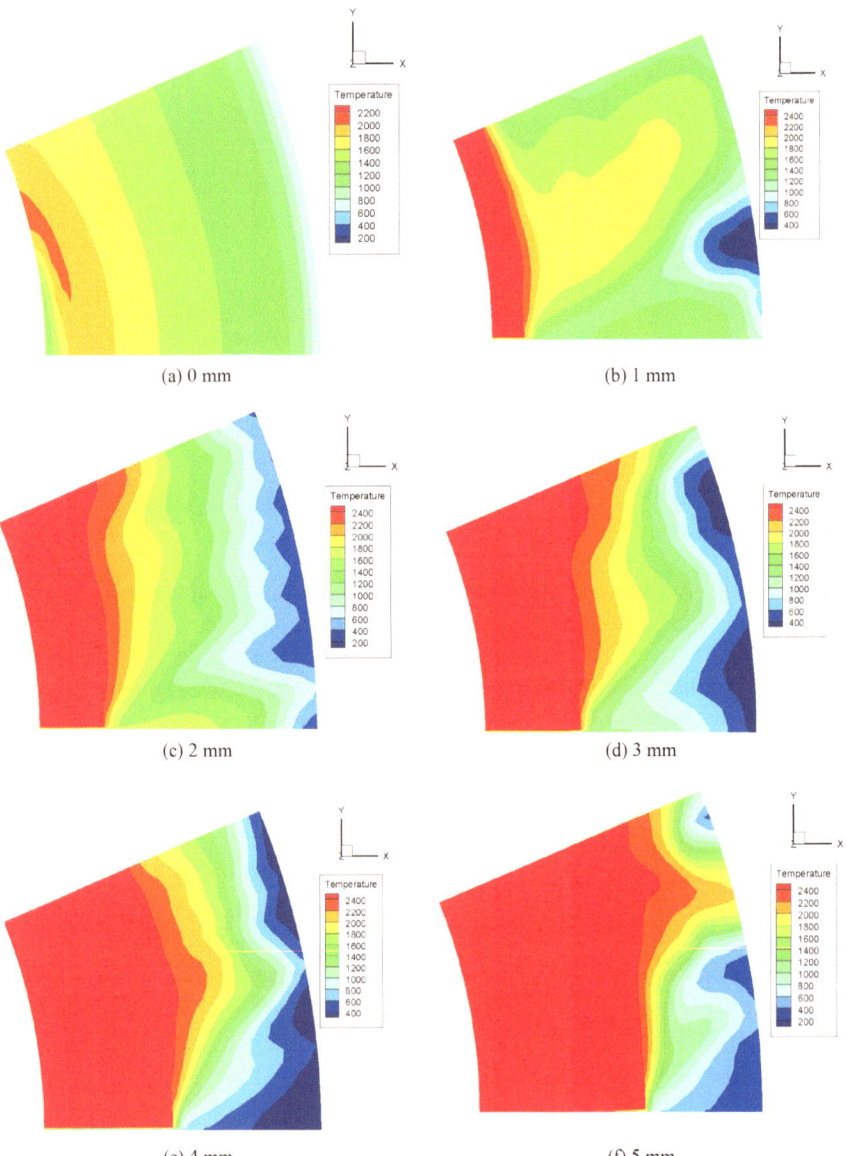

Fig. 5.11 Comparison of the temperature distribution at the symmetric plane of the heat exchanger core under different control runner lengths (unit: K)

Table 5.1 Comparison of laminate parameters with different control runner lengths

Length (mm)	Temperature (K)	Speed (m/s)	Total temperature (K)
1	2162	845	2186
2	2192	871	2163
3	2137	867	2163
4	1962	909	1990
5	1970	795	1993

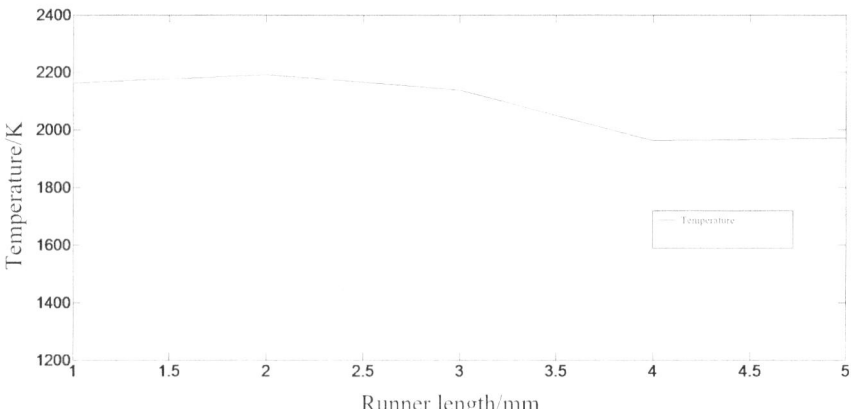

Fig. 5.12 Comparison of the average temperature and average total temperature at the outlet of the heat exchanger core under different control runner lengths

speed of the high-temperature part of the dispersion area is lower than that of the mainstream part, the flow is not stationary at the outlet of the heat exchanger core, with a speed above 400 m/s. Figure 5.15 compares the pressure distribution on the symmetric plane of the heat exchanger core under different control runner lengths. The pressure drops of several designs increase with the increase in the control runner length, and the pressure drops are all concentrated along the length direction of the control runner. When the length increases, to ensure a consistent thrust, the inlet pressure of the entire system needs to be increased. Of course, the high speed in the flow area is mainly caused by the high pressure drop. In actual operation, the selection of an appropriate pressure drop depends on the needed thrust. In this book, the heating effect of different designs is only compared under the same pressure drop.

Therefore, based on comprehensive consideration, a control runner length of 2–3 mm is optimal among the several design schemes. The average temperature of the propellant at the laminate outlet is 2192 K.

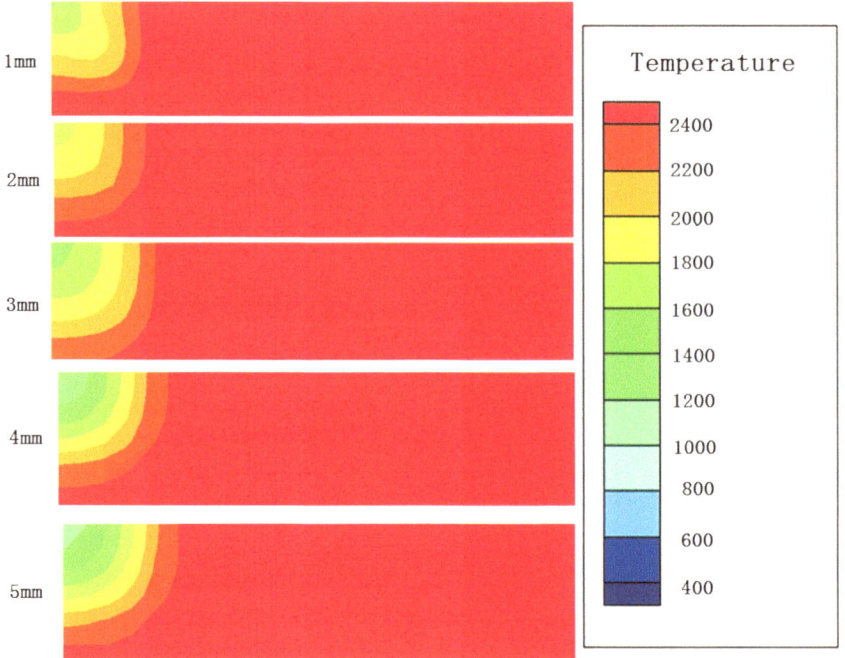

Fig. 5.13 Distribution of temperature at the cross-section of the heat exchanger core outlet under different control runner lengths

5.3.1.2   Temperature Distribution Characteristics of the Solid Region

Figure 5.16 shows the temperature distribution in the solid region of the laminate with different control runner lengths, and the temperature exhibits a linear decreasing pattern from the inside (close to the absorption cavity) to the outside. A comparison of the temperature at the outer low temperature end shows that under the same heating conditions, the temperature minimum area of the laminate solid is large when the control runner length is 2 mm, and the temperature minimum area is smallest when the control runner length is 4 mm, suggesting that the cooling effect of the laminate structure with 2 mm control runners is better, and the propellant obtains the most energy from the laminates.

5.3.2   *Effect of Control Runner Cross-Sectional Area on Heat Transfer of Laminates*

The cross-sectional area of the control runner is also an important factor affecting the heating effect of the laminated heat exchanger core. Taking a control runner

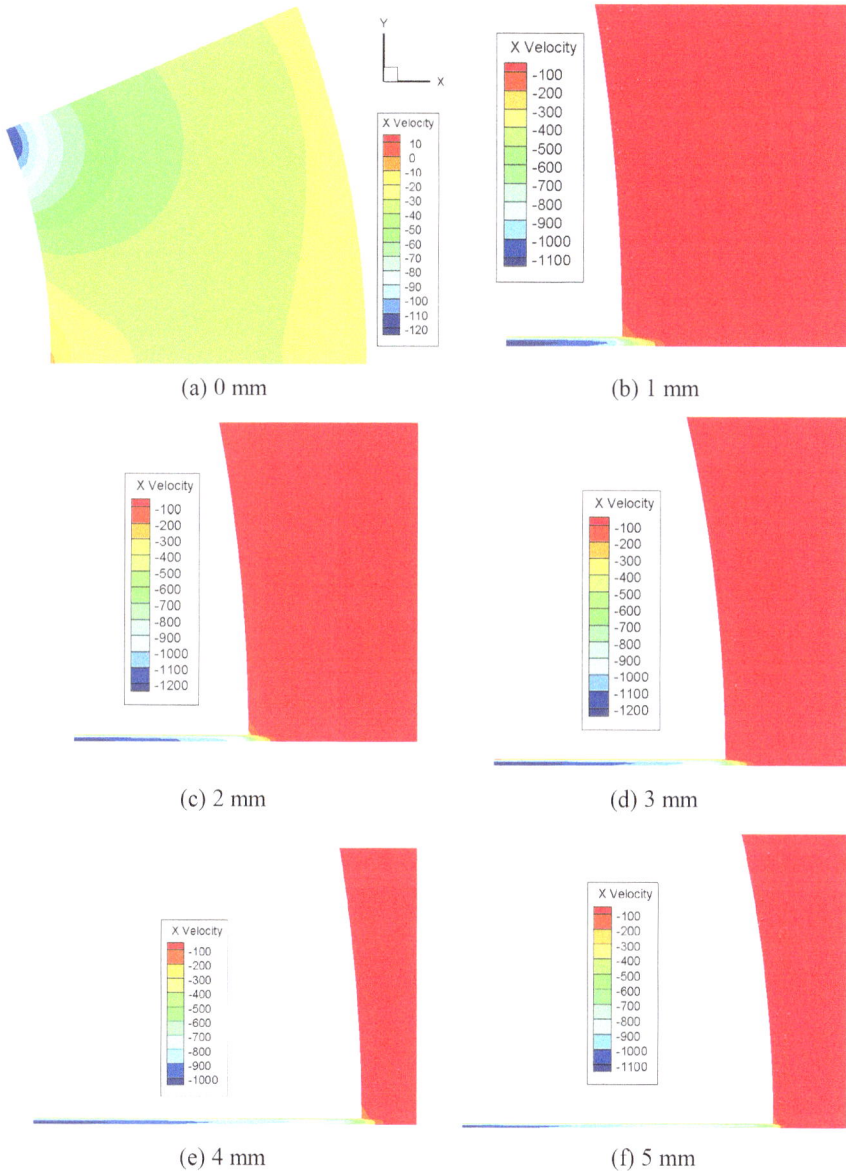

Fig. 5.14 Comparison of the speed distributions at the symmetric plane of the heat exchanger core under different control runner lengths

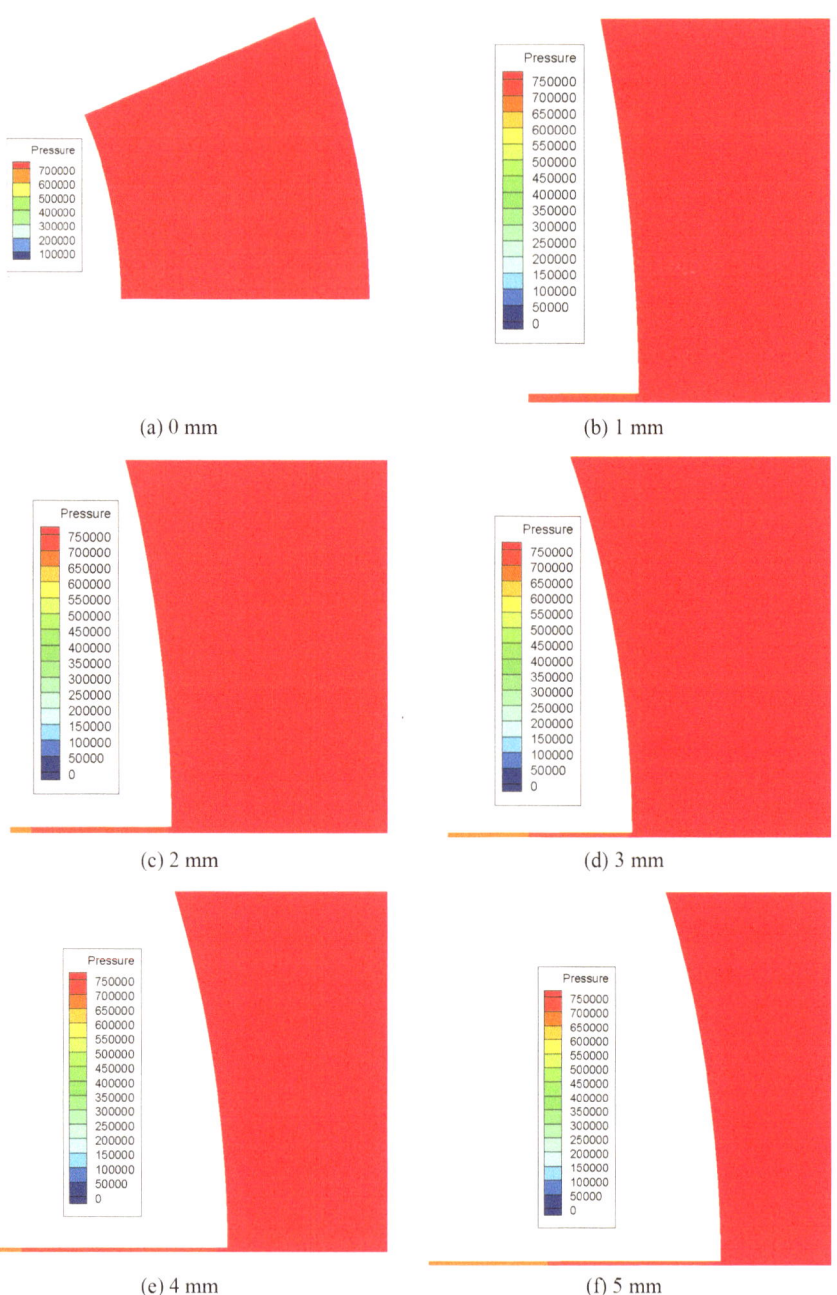

(a) 0 mm   (b) 1 mm

(c) 2 mm   (d) 3 mm

(e) 4 mm   (f) 5 mm

Fig. 5.15 Comparison of pressure distribution on the symmetric plane of the heat exchanger core under different control runner lengths

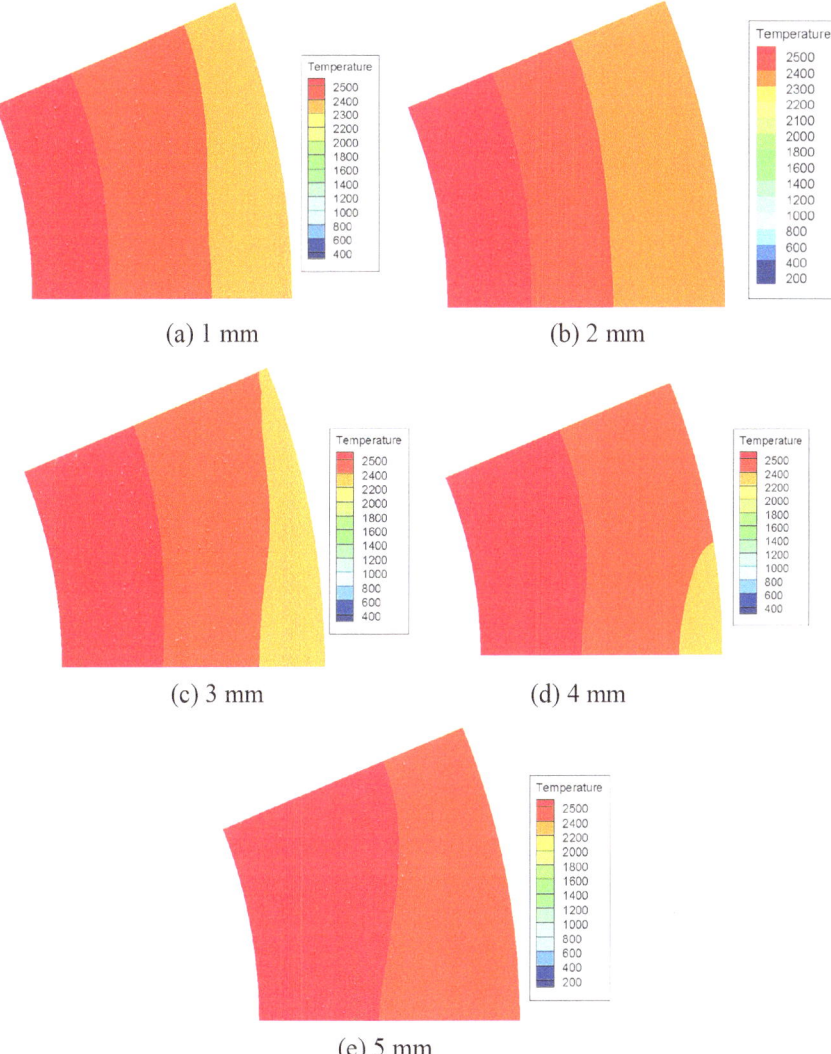

(a) 1 mm

(b) 2 mm

(c) 3 mm

(d) 4 mm

(e) 5 mm

Fig. 5.16 Comparison of the temperature distribution at the symmetric plane in the solid area of the heat exchanger core under different control runner lengths

length of 2.5 mm as an example, cases with the control runner cross-sectional areas of 0.01, 0.02 and 0.03 mm^2 are analyzed. The distributions of the temperature at the symmetric plane of the three cases are shown in Fig. 5.17.

The symmetric plane with the lowest average temperature is selected for comparison. The temperature at other locations is higher, and the temperature of the near wall at the outlet is above 2300 K. A comparison of the temperature distribution cloud map of the symmetric plane shows that in the dispersion region, the 0.01 mm^2

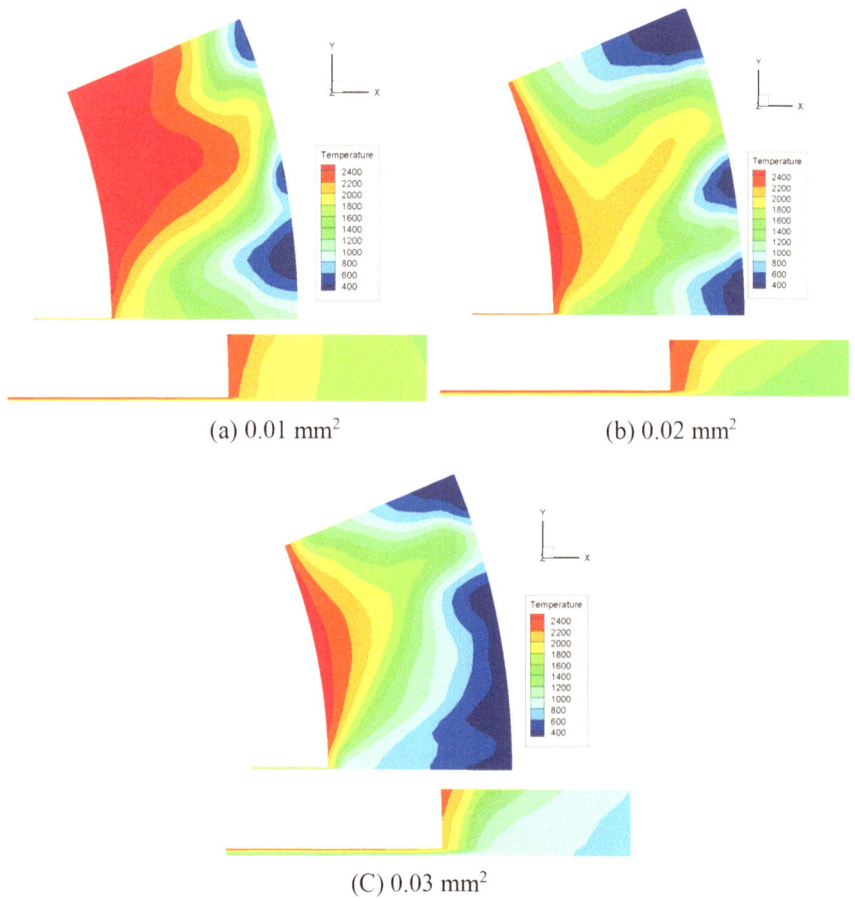

(a) 0.01 mm^2 (b) 0.02 mm^2

(C) 0.03 mm^2

Fig. 5.17 Comparison of the cross-sectional areas of the control runners in the laminates

design has the best heating effect, and the heating temperature at the runner outlet can exceed 2300 K. Under the same working conditions, for the 0.02 mm^2 and 0.03 mm^2 runner designs, the temperature is only approximately 2200 K and 2000 K, respectively. The laminate outlet temperatures of the three designs are compared in Table 5.2. With the increase in the cross-section of the control runner, the heating effect of the heat exchanger core on the working fluid is reduced. The control runner with a cross-sectional area of 0.01 mm^2 has the best heating effect among the three designs, followed by the control runner with a cross-sectional area of 0.02 mm^2. When considering the processing cost and difficulty, a finer control runner is no longer desirable, and a runner of 0.02 mm^2 is chosen as the final design; that is, a circular control runner with a radius of 0.08 mm is chosen.

Table 5.2 Comparison of the laminate outlet temperatures under different control runner cross-sectional areas

Cross-sectional area (mm^2)	Temperature (K)	Total temperature (K)
0.01	2356	2402
0.02	2279	2321
0.03	1980	2016

Chapter 6
Simulation of the Heating of Laminated Heat Exchanger Core Under Variable Operating Conditions

6.1 Introduction

The laminated heat exchange core is the main part that heats the propellant/working fluid in a solar thermal propulsion (STP) system. In this chapter, through a simulation and analysis of the working process of a heat exchanger core with a laminate structure, the laminate heating characteristics of the STP system under variable operating conditions are obtained.

6.2 Physical Model and Calculation Method for the Laminate

6.2.1 Physical Model

The performance of a solar thermal propulsion (STP) system under variable operating conditions is largely determined by the high-efficiency heat exchange microchannels. The heat transfer performance of the laminates directly affects the thermal performance parameters of the gaseous working fluid before entering the nozzle.

The heat exchanger core of a solar thermal thruster under variable working conditions uses a laminated heat exchange microchannel structure, and the heat exchange area between the propellant/working fluid and the heat exchanger core wall is increased through shunting by multiple runners, fully heating the propellant/gaseous working fluid in the heat exchanger core, which improves the convective heat transfer efficiency of the propellant/working fluid in the heat exchange microchannel. Figure 6.1 shows the structure of a solar thermal thruster under variable working conditions. The heat exchanger core structure is described in Sect. 5.2.1.

© National University of Defense Technology Press 2025
M. Huang et al., *Solar Thermal Thruster*, https://doi.org/10.1007/978-981-97-7490-6_6

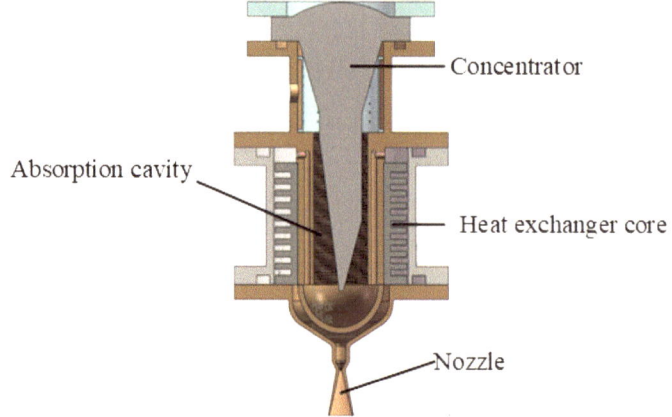

Fig. 6.1 Solar thermal thruster under variable working conditions

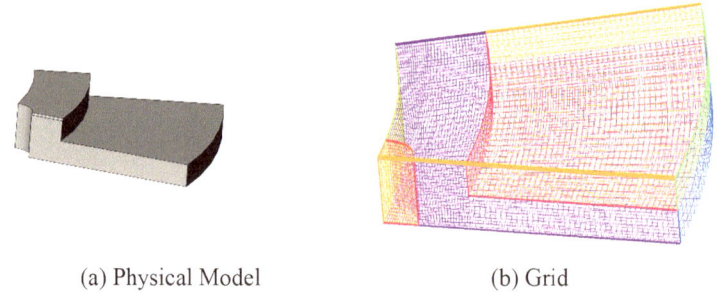

(a) Physical Model   (b) Grid

Fig. 6.2 3D grid of circular heat exchange microchannels

Figure 6.2a and b show the results of the laminate model obtained by using a structural grid and considering the actual shape of the heat exchange microchannel and the effect of the vertical interlayer.

6.2.2 *Control Equations*

In a cylindrical coordinate system, the 3D NS equation in the conservation form can be written as:

$$\frac{\partial Q}{\partial t} + \frac{\partial E}{\partial x} + \frac{\partial F}{\partial y} + \frac{\partial G}{\partial z} = \frac{\partial E_v}{\partial x} + \frac{\partial F_v}{\partial y} + \frac{\partial G_v}{\partial z} \tag{6.1}$$

where

$$Q = \begin{bmatrix} \rho \\ \rho u \\ \rho v_\theta \\ \rho v_r \\ \rho E \end{bmatrix} \quad E = \begin{bmatrix} \rho u \\ P + \rho u u \\ \rho v_\theta u \\ \rho v_r u \\ \rho H u \end{bmatrix} \quad F = \begin{bmatrix} \rho v_\theta \\ \rho v_\theta u \\ P + \rho v_\theta v_\theta \\ \rho v_\theta v_r \\ \rho H v_\theta \end{bmatrix} \quad G = \begin{bmatrix} \rho v_r \\ \rho v_r u \\ \rho v_\theta v_r \\ P + \rho v_r v_r \\ \rho H v_r \end{bmatrix}$$

$$E_v = \begin{bmatrix} 0 \\ \tau_{xx} \\ \tau_{\theta x} \\ \tau_{rx} \\ u\tau_{xx} + v_\theta \tau_{\theta x} + v_r \tau_{xx} - q_x \end{bmatrix} \quad F_v = \begin{bmatrix} 0 \\ \tau_{x\theta} \\ \tau_{\theta\theta} \\ \tau_{r\theta} \\ u\tau_{x\theta} + v_\theta \tau_{\theta\theta} + v_r \tau_{r\theta} - q_\theta \end{bmatrix}$$

$$G_v = \begin{bmatrix} 0 \\ \tau_{xr} \\ \tau_{\theta r} \\ \tau_{rr} \\ u\tau_{xr} + v_\theta \tau_{\theta r} + v_r \tau_{rr} - q_r \end{bmatrix}$$

where u, v_θ, and v_r represent the speed in the z axial, circumferential θ and radial r directions, respectively, and e is the internal energy contained in the unit mass of fluid. Then, the total energy contained in the unit mass of fluid is $E = e + \frac{1}{2}(u^2 + v_\theta^2 + v_r^2)$, P is the fluid pressure, and ρ is the fluid density.

In each vector, the shear stress τ_{ij} and heat transfer term q_i are specifically expressed as follows:

$$\tau_{xx} = \mu\left(2\frac{\partial u}{\partial x} - \frac{2}{3}\nabla\cdot\vec{V}\right) = \frac{2}{3}\mu\left(2\frac{\partial u}{\partial x} - \frac{\partial rv_r}{r\partial r} - \frac{\partial v_\theta}{r\partial\theta}\right) \tag{6.2}$$

$$\tau_{rr} = 2\mu\frac{\partial v_r}{\partial r} - \mu\frac{2}{3}\nabla\cdot\vec{V} = \frac{2}{3}\mu\left(2\frac{\partial rv_r}{r\partial r} - \frac{\partial v_\theta}{r\partial\theta} - \frac{\partial u}{\partial x}\right) - 2\mu\frac{v_r}{r} \tag{6.3}$$

$$\tau_{\theta\theta} = 2\mu\frac{1}{r}\left(\frac{\partial v_\theta}{\partial\theta} + v_r\right) - \frac{2}{3}\mu\nabla\cdot\vec{V} = \frac{2}{3}\mu\left(2\frac{\partial v_\theta}{r\partial\theta} - \frac{\partial rv_r}{r\partial r} - \frac{\partial u}{\partial x}\right) + \mu\frac{2v_r}{r} \tag{6.4}$$

$$\tau_{x\theta} = \tau_{\theta x} = \mu\left(\frac{\partial v_\theta}{\partial x} + \frac{\partial u}{r\partial\theta}\right) \tag{6.5}$$

$$\tau_{xr} = \tau_{rx} = \mu\left(\frac{\partial v_r}{\partial x} + \frac{\partial u}{\partial r}\right) = \mu\left(\frac{\partial v_r}{\partial x} + \frac{\partial ru}{r\partial r} - \frac{u}{r}\right) \tag{6.6}$$

$$\tau_{r\theta} = \tau_{\theta r} = \mu\left(\frac{\partial v_\theta}{\partial r} + \frac{\partial v_r}{r\partial\theta} - \frac{v_\theta}{r}\right) = \mu\left(\frac{\partial rv_\theta}{r\partial r} + \frac{\partial v_r}{r\partial\theta}\right) - 2\mu\frac{v_\theta}{r} \tag{6.7}$$

$$\nabla\cdot\vec{V} = \frac{\partial u}{\partial x} + \frac{\partial rv_r}{r\partial r} + \frac{\partial v_\theta}{r\partial\theta} \tag{6.8}$$

$$q_z = -k\frac{\partial T}{\partial z}, \; q_\theta = k\frac{\partial T}{\partial \theta}, \; q_r = k\frac{\partial T}{\partial r} \tag{6.9}$$

The differential equation of the energy transport equation of the Fourier law of heat transfer in the solid is:

$$\frac{\partial^2 T}{\partial x_j \partial x_j} = 0 \tag{6.10}$$

where T is the solid temperature.

The energy equilibrium equation based on the finite element method is:

$$KT = Q \tag{6.11}$$

where K is the transfer matrix, T is the nodal temperature vector, and Q is the nodal heat flow rate vector.

The variation pattern of the solar radiation intensity versus the distance is shown in the following equation [108–110].

$$I_{\lambda,L} = I_{\lambda,0} \exp\left[-\int_0^L \beta_\lambda(y)dy\right] \tag{6.12}$$

where y is the propagation direction of light, $I_{\lambda,L}$ is the radiation intensity at $y = L$, and $I_{\lambda,0}$ is the radiation intensity at $y = 0$.

β_λ is the spectral attenuation coefficient:

$$\beta_\lambda(x) = \kappa_\lambda(x) + \sigma_{s\lambda}(x) \tag{6.13}$$

where k_λ is the spectral absorption coefficient and $\sigma_{s\lambda}$ is the spectral scattering coefficient.

Using the standard E_α dual-equation turbulence model, the basic transport equation is obtained as follows:

$$\frac{\partial(\rho\varepsilon)}{\partial t} + \frac{\partial(\rho\varepsilon u_i)}{\partial x_i} = \frac{\partial}{\partial x_j}\left[\left(\mu + \frac{\mu_i}{\alpha_k}\right)\frac{\partial \varepsilon}{\partial x_j}\right]$$
$$+ B_{1\varepsilon}\frac{\varepsilon}{K}(T_k + B_{3\varepsilon}T_b) - B_{2\varepsilon}\rho\frac{\varepsilon^2}{k} + S_\varepsilon \tag{6.14}$$

$$\frac{\partial(\rho k)}{\partial t} + \frac{\partial(\rho k u_i)}{\partial x_i} = \frac{\partial}{\partial x_j}\left[\left(\mu + \frac{\mu_i}{\alpha_k}\right)\frac{\partial k}{\partial x_j}\right]$$
$$+ T_k + T_b - \rho\varepsilon - X_M + S_k \tag{6.15}$$

where T_k and T_b are the turbulent kinetic energy generation term and the generation term of turbulent kinetic energy due to buoyancy, respectively; X_M is the pulsation term in turbulent flow; S_k and S_ε are user-defined source terms; α_k and α_ε are the Prandtl numbers corresponding to the turbulent kinetic energy k and dissipation rate ε; and $B_{1\varepsilon}$, $B_{2\varepsilon}$ and $B_{3\varepsilon}$ are empirical constants.

6.2.3 Boundary Conditions

The inner wall of the heat exchanger core is in close contact with the wall of the absorption cavity, and the temperature in the absorption cavity can exceed 2300 K, so for the heat exchanger core, the boundary condition of the inner wall of the laminate is set to 2400 K.

The propellant inlet conditions are obtained based on the calculation results for each working condition in Sect. 3.6.3, as shown in Table 6.1. The inlet pressure condition is set to 0.4 MPa.

6.3 Temperature Distribution Characteristics of the Laminates

Figure 6.3 shows the temperature distributions of the 1/8 laminate and working fluid obtained from the fluid–solid coupling heat transfer numerical simulation. The temperature of the fluid part is heated from the initial 300–1800 K, and the temperature of the solid part decreases from 2400 (at the inner side) to 1800 K, showing a clear change. The simulation results show that the cross-sectional area and shape of the heat exchange microchannel have little impact on the heating effect, and the temperature of the working fluid at the center of the runner outlet after heating is approximately 1800 K.

The total depth of the heat exchange microchannels in the calculation model is 0.08 mm. Figure 6.4 shows the temperature distribution of the heat exchange microchannels. The temperature distribution of the working fluid shows a gradual rising trend from the inlet to the outlet of the channel. In addition, the temperature distribution gradually increases from the center of the runner to the laminate wall. The heat exchange microchannel is the main place where the working fluid is heated, but the temperature gradient in the channel is relatively small; the working fluid is rapidly

Table 6.1 Boundary conditions for propellant inlets

Working condition	0.1 N	0.2 N	0.3 N	0.4 N	0.5 N
Mass flow rate ($\times 10^5$ kg/s)	1.178	2.356	3.534	4.713	5.891
Pressure (MPa)	0.6	0.6	0.6	0.6	0.6

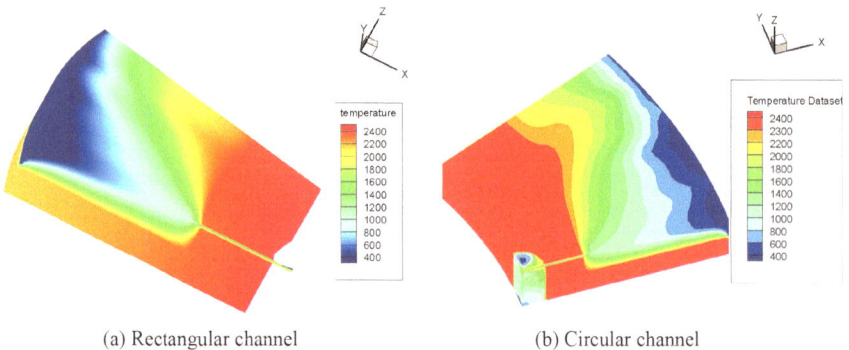

(a) Rectangular channel   (b) Circular channel

Fig. 6.3 Temperature distribution of the laminates and working fluid

heated to approximately 1400 K after entering the runner. As shown in Fig. 6.4a, the temperature of the working fluid in the center of the symmetric plane of the runner is increased to approximately 1800 K at the laminate outlet, and the temperature of the working fluid in the runner near the laminate also exceeds 2200 K. The working fluid temperature distribution of a heat exchange microchannel with a depth of 0.04 mm is shown in Fig. 6.4b. The temperature of the working fluid in the runner near the laminate wall is close to 2300 K, and the overall temperature distribution is similar to that of the symmetric plane. Figure 6.4c shows the temperature distribution in the heat exchange microchannel near the laminate wall, and the temperature of the working fluid at the runner outlet basically reaches approximately 2300 K.

Divided along the circumferential direction of the laminates, the temperature distribution of the solid part inside the laminated heat exchanger core is shown in Fig. 6.5. The temperature at the inner side of the heat exchanger core (i.e., working fluid inlet) is approximately 2400 K, and the temperature at the outer side (i.e., working fluid outlet) is approximately 1800 K. In general, the temperature distribution of the entire heat exchanger core varies from 1800 to 2400 K. The fluid–solid coupling method is used in the simulation. Due to the restriction of the mutual influence between the fluid and the solid, the temperature of the solid part of the laminate exhibits a gradient change with the fluid flow. For the temperature of the solid part of the laminate, the temperature of the cold end of the solid is 1800 K, and the temperature of the hot end of the solid is 2400 K, indicating that the cooling effect of the working fluid on the solid part is superior, and the working fluid is heated to a higher temperature.

Figure 6.6 shows the fluid speed distributions obtained from the cross-sections extracted at different depths of the laminated heat exchange microchannels. Figure 6.6 shows that before the working fluid enters the heat exchange microchannel, the flow speed of the working fluid is approximately 500 m/s and does not change significantly, the flow speed of the working fluid at the center of the heat exchange microchannel is the highest, the flow speed at the symmetric plane can reach a maximum of 2500 m/s, the flow speed near the wall is approximately 1000 m/s, and

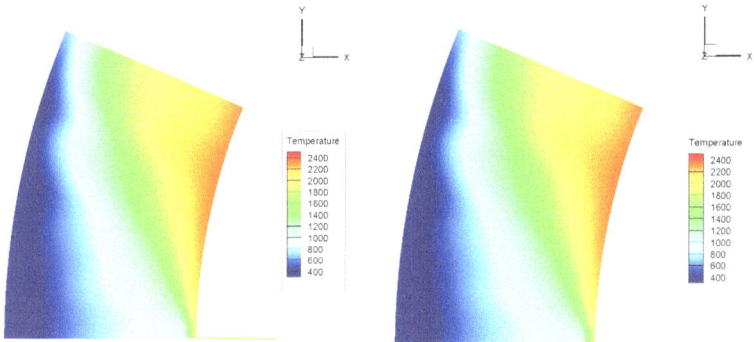

(a) Symmetric plane of the heat exchange microchannel (b) The microchannel depth = 0.04 mm

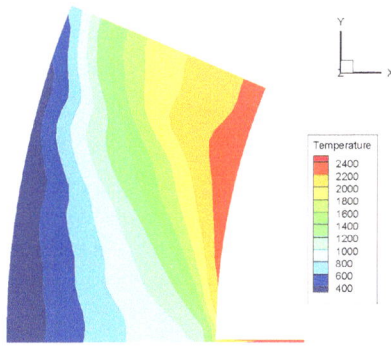

(c) The microchannel depth = 0.08 mm

Fig. 6.4 Cross-sectional temperature distribution of the fluid part inside the laminated heat exchanger core

Fig. 6.5 Cross-sectional temperature distribution of the solid part inside the laminated heat exchanger core

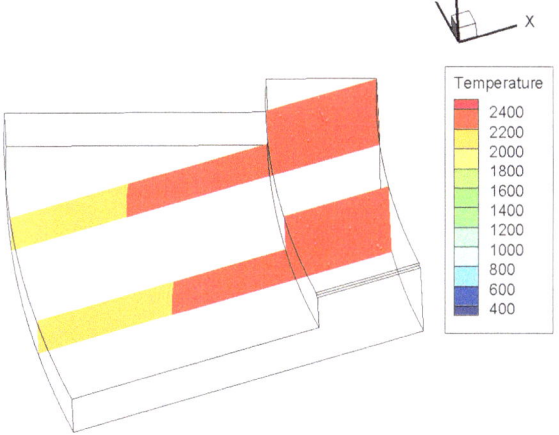

the speed of the working fluid at the center of the channel decreases toward the wall, which is due to the existence of the fluid boundary layer.

Figure 6.7 shows the pressure distribution in the heat exchange microchannels at the symmetric plane of the runners in the laminated heat exchanger core. As shown in Fig. 6.7, after passing through the heat exchange microchannel, the fluid temperature and speed are greatly increased, and the fluid pressure gradually decreases from 0.4 MPa at the inlet to approximately 0.3 MPa at the outlet of the heat exchange microchannel. The pressure drop in the heat exchanger core is approximately 0.1 MPa, which has little impact on the performance of the thruster under variable working conditions. Therefore, when considering changing the thrust of the thruster, the propellant inlet pressure can be maintained at 0.4 MPa, and the thrust can be changed only by changing the solar radiation energy and the propellant flow rate.

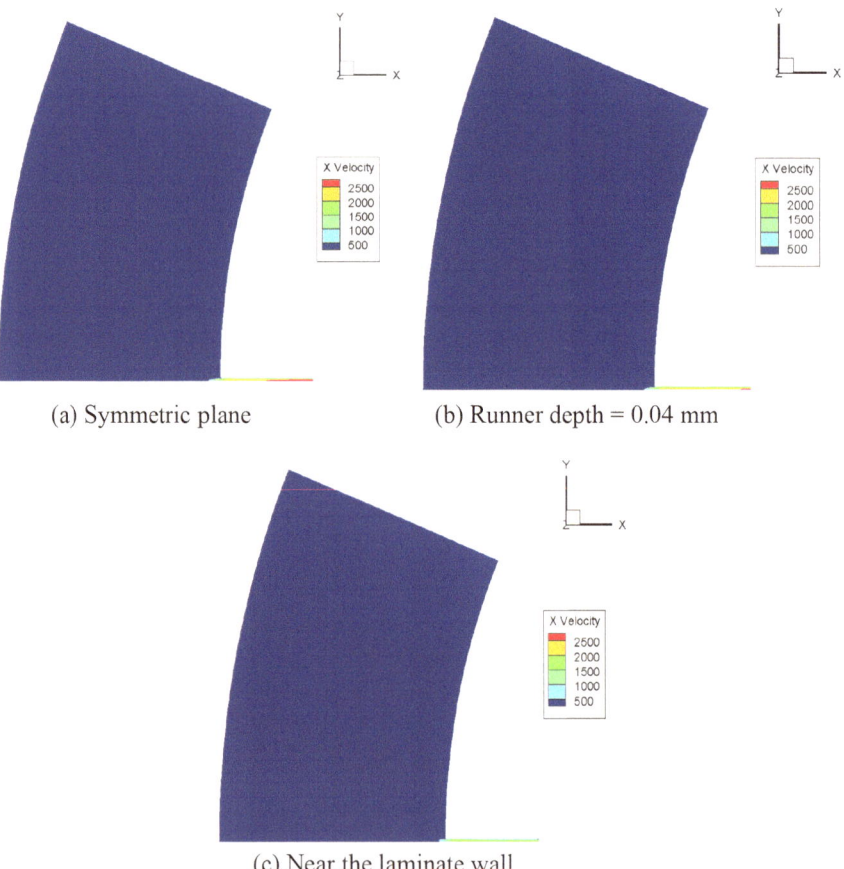

(a) Symmetric plane   (b) Runner depth = 0.04 mm

(c) Near the laminate wall

Fig. 6.6   Working fluid speed distribution in the cross-sections at different depths

Fig. 6.7 Fluid pressure distribution in the laminated microchannels

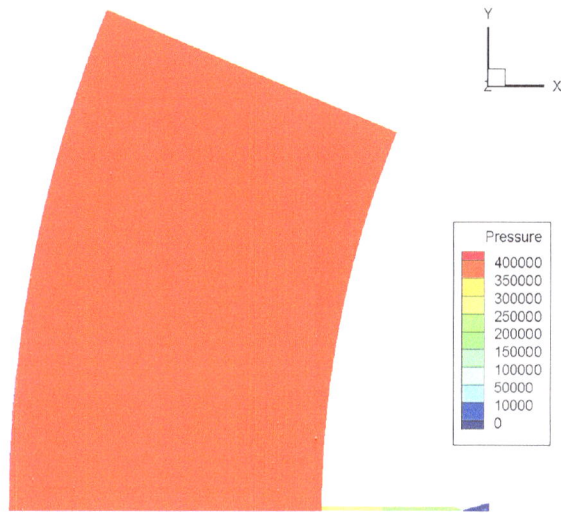

6.4 Effect of Laminated Structure Parameters on the Heating Effect Under Variable Working Conditions

6.4.1 Runner Cross-Sectional Area

The cross-sectional area of the heat exchange microchannels of the heat exchanger core affects the heating of the working fluid in the runner, which has a certain impact on the flow speed and temperature of the working fluid. Simulation analysis of the working fluid temperature distribution pattern with different cross-sectional areas of the heat exchange microchannel was performed, with the length of the heat exchange microchannel being 2.5 mm. Figure 6.8 shows the temperature distribution cloud map on the symmetric plane of the working fluid under different situations.

An analysis of the working fluid temperature distribution pattern in the runners with different cross-sectional areas shows that in the heat exchange microchannel, the 0.01 mm² runner has the best heating effect, and temperature at the runner outlet can exceed 2300 K. Under the same working conditions, for runners of 0.02 mm² and 0.03 mm², the temperature can only reach approximately 2200 K and 2000 K, respectively. The simulation results show that the smaller the cross-sectional area, the higher the temperature of the working fluid after heating, the greater the internal energy obtained by the working fluid, the greater the kinetic energy obtained by the thruster, and the better the performance.

An integration mean of cross-sectional temperature distribution at the outlet of a laminated runner with different cross-sectional areas is performed. Table 6.2 shows a comparison of the temperatures at the outlet of a laminated runner with different cross-sectional areas. With the increase in the cross-sectional area of the heat

Fig. 6.8 Temperature
comparison of the laminated
heat exchange
microchannels with different
cross-sectional areas

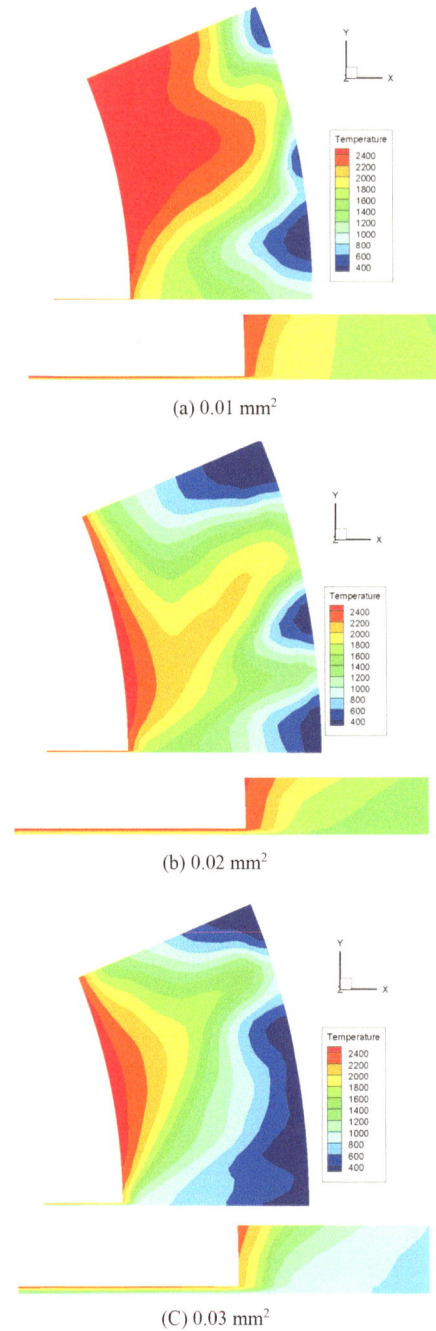

(a) 0.01 mm^2

(b) 0.02 mm^2

(C) 0.03 mm^2

Table 6.2 Comparison of temperatures at the exit of the layers with different cross-sectional areas of the flow passages

Cross-sectional area (mm^2)	Temperature (K)	Total temperature (K)
0.01	2356	2402
0.02	2279	2321
0.03	1980	2016

exchange microchannel, the heating effect of the heat exchanger core on the working fluid decreases. The simulation results show that the heat exchange microchannel with a cross-sectional area of 0.01 mm^2 has the best heating effect, followed by the heat exchange microchannel with a cross-sectional area of 0.02 mm^2, and the heat exchange microchannel with a cross-sectional area of 0.03 mm^2 has the worst heating effect. Based on the economic cost of thruster processing and the difficulty of aperture processing in engineering, a circular heat exchange microchannel with a cross-sectional area of 0.02 mm^2 is selected.

6.4.2 Effect of Runner Length

The structural parameters of the laminated heat exchanger core affect the heating effect of the laminates to a certain extent. The simulation studies the influences of the lengths and cross-sectional areas of the heat exchange microchannels on the heating effect.

For heat exchange microchannels of different lengths, the numerical simulation obtains the working fluid temperature distribution cloud map on the symmetric plane of the laminate, as shown in Fig. 6.9. The heat exchange microchannels have a significant effect on the heating of working fluid. When only parallel laminates are used to heat the working fluid, the heating effect is very poor, and the temperature of the working fluid after heating is only approximately 1000 K. After heat exchange microchannels are used, the working fluid can be heated to above 1800 K.

By performing area integration on the cross-section at the runner outlet, the average temperature and average total temperature of the cross-section at the outlet are shown in Table 6.3. With increasing runner length, the temperature of the cross-section at the outlet first increases and then decreases and then increases; the working fluid speed of the cross-section at the outlet also shows a pattern of increasing–decreasing–increasing. The change in total temperature is relatively monotonic, and the total temperature gradually decreases with increasing length. Comprehensive analysis shows that when the heat exchange microchannel length is 2–3 mm, the total temperature at the outlet of the runner is relatively high, and the flow speed is relatively high, which is beneficial for improving the overall performance of the thruster.

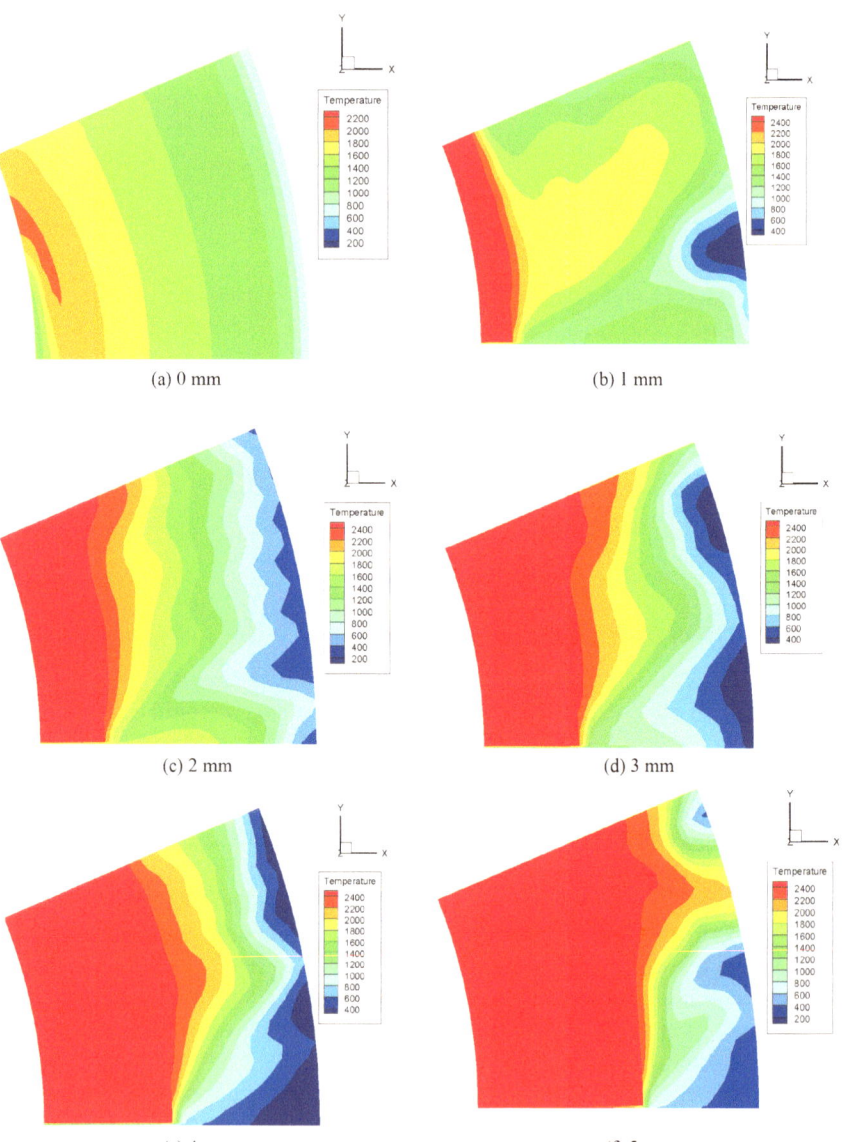

(a) 0 mm

(b) 1 mm

(c) 2 mm

(d) 3 mm

(e) 4 mm

(f) 5 mm

Fig. 6.9 Temperature distribution pattern of the working fluid at the symmetric plane with different runner lengths

Table 6.3 Comparison of the outlet temperature and flow speed of the laminated heat exchange microchannel with different lengths

Length (mm)	Temperature (K)	Speed (m/s)	Total temperature (K)
1	2162	845	2186
2	2192	871	2163
3	2137	867	2163
4	1962	909	1990
5	1970	795	1993

Figure 6.10 shows the trend of the average temperature and the average total temperature at the outlet of the heat exchanger core with different heat exchange microchannel lengths. After entering the heat exchange microchannel, the working fluid is rapidly heated to a relatively high temperature. The longer the channel is, the larger the flow speed of the working fluid is in the channel, but the heating effect can be reduced. Considering the overall quality requirement and structural design of the thruster, a 2–3 mm long heat exchange microchannel is preferred.

The temperature distributions at the cross-sections at the outlet of heat exchanger core under different heat exchange microchannel lengths are shown in Fig. 6.11. As the length of the heat exchange microchannel increases, the proportion of the high-temperature area at the cross-section at the outlet of the channel also gradually increases. This is because the longer the heat exchange microchannel is, the more strongly the propellant is compressed before entering the runner, and the larger the temperature gradient is. Once the runner length exceeds 3 mm, there is a trend of a

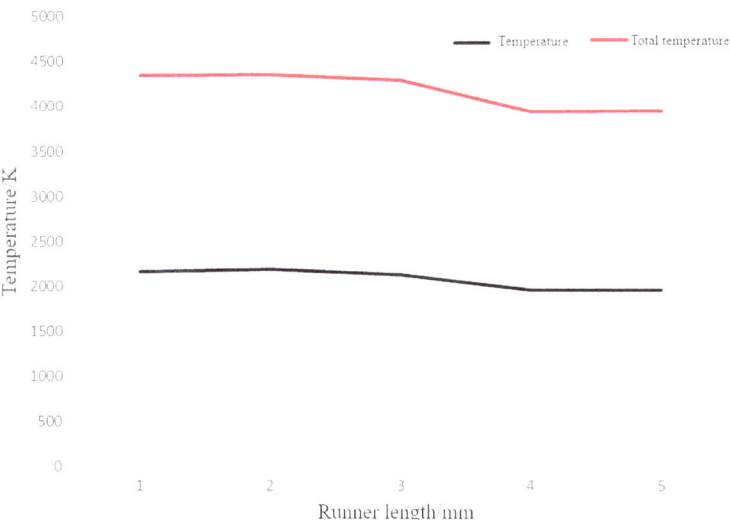

Fig. 6.10 Comparison of the average temperature and average total temperature at the outlet of the heat exchanger core under different heat exchange microchannel lengths

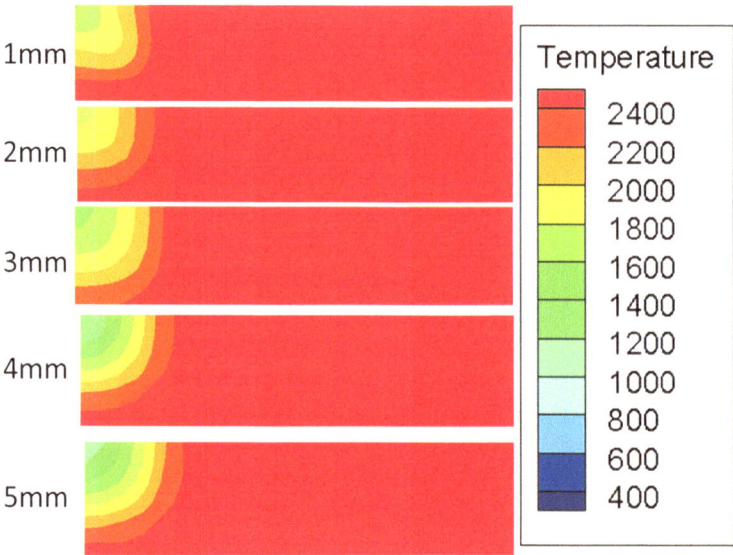

Fig. 6.11 Temperature distribution at the cross-section of the outlet of the heat exchanger core under different control runner lengths

proportional low-temperature area of the propellant increasing and a proportional high-temperature area of the propellant decreasing at the outlet, suggesting that continued lengthening of the heat exchange microchannel could reduce the heating effect on the propellant. The comparison of the speed distribution of the working fluid at the symmetric plane of the heat exchanger core under different heat exchange microchannel lengths is shown in Fig. 6.12. Figure 6.12 shows that although the speed of the high-temperature part at the outlet is slower than that of the mainstream part, the flow at the outlet of the heat exchanger core is not stationary, with a minimum speed of 500 m/s.

Therefore, using 2–3 mm as the length of the heat exchange microchannels is optimal for the heating effect. The average temperature of the propellant at the laminate outlet is 2192 K.

Figure 6.13 shows the temperature distribution of the solid part of the laminate obtained from the simulation analysis of the heat exchange microchannels with different lengths. In Fig. 6.13, the temperature distribution of the solid part of the laminate shows a gradually decreasing trend from the inside to the outside. The larger the temperature difference is between the inside and the outside of the solid part, the more the heat is transferred by the working fluid, and the higher the heat transfer efficiency of the laminates is, that is, the higher the working fluid temperature is. The design of 2 mm heat exchange microchannels has the best heating effect when the total length of the laminated runner is approximately 8 mm, so when the heat exchange microchannel is one-quarter of the total length, the working fluid obtains the most energy from the laminates.

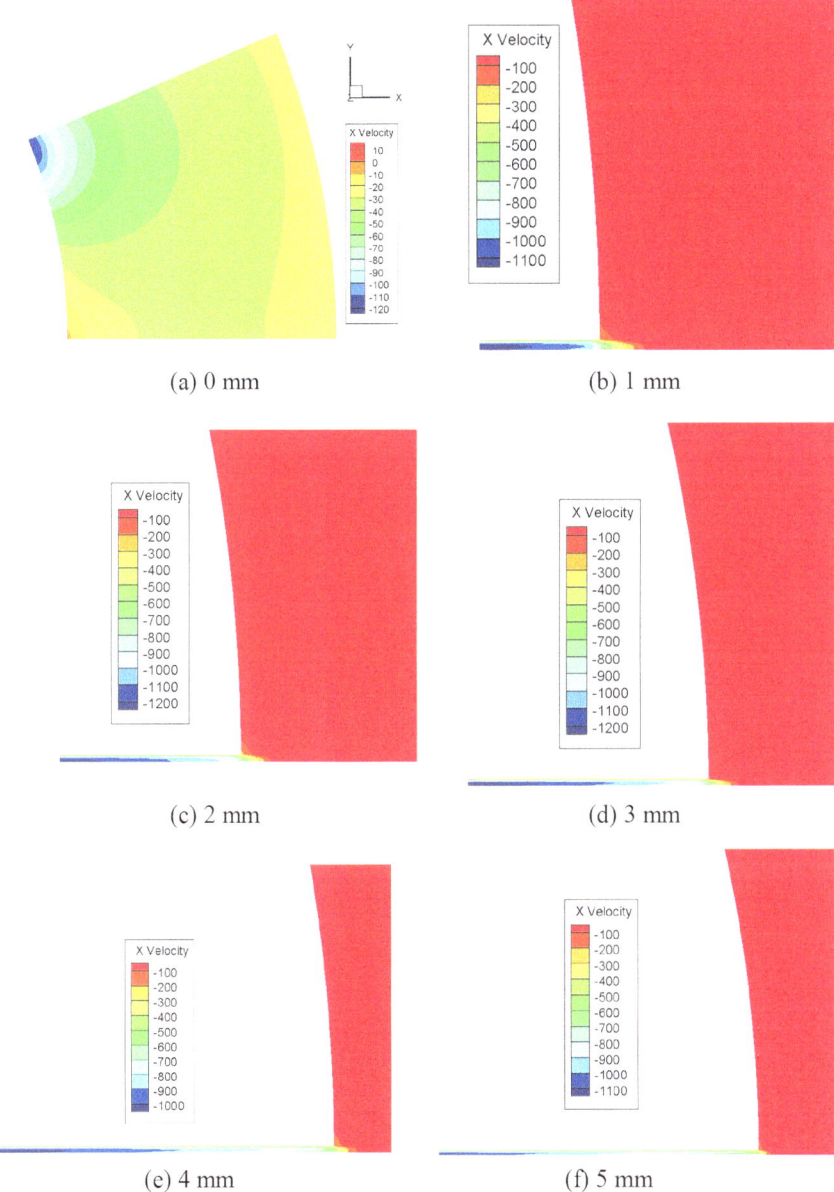

Fig. 6.12 Comparison of working fluid speed distribution on the symmetric plane of heat exchanger core under different heat exchange microchannel lengths

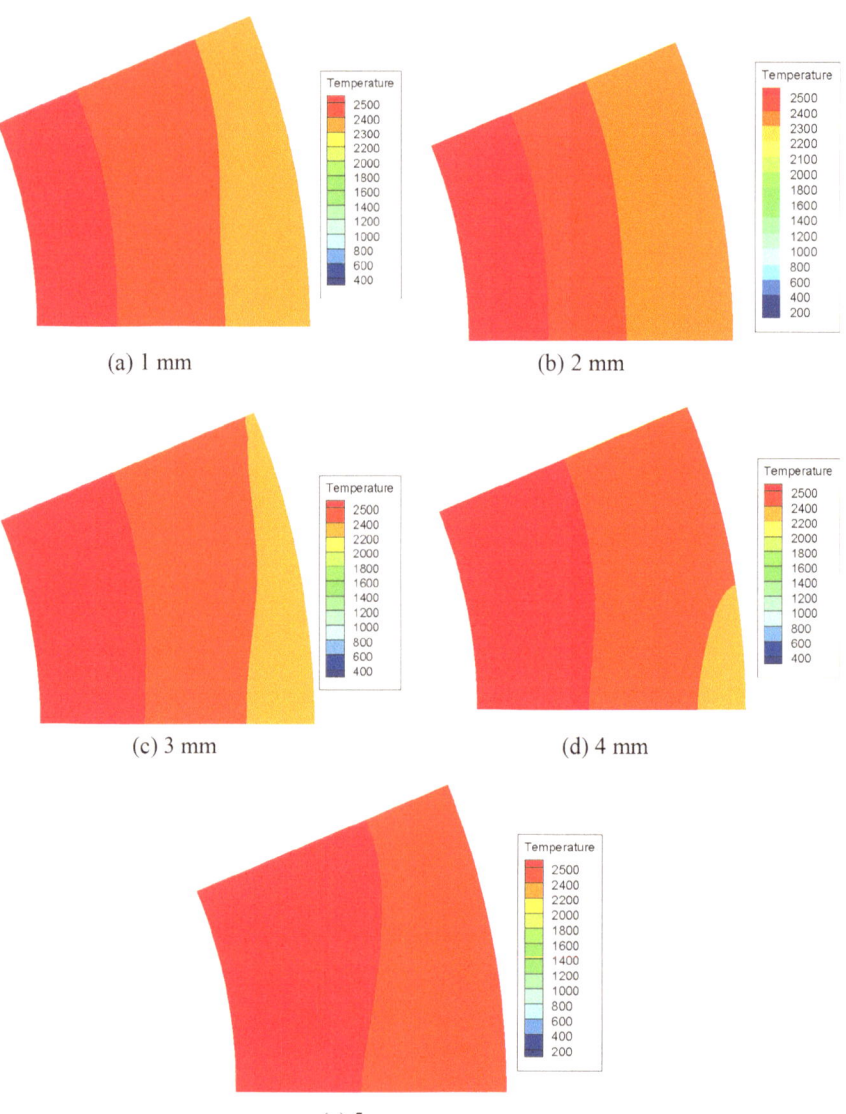

Fig. 6.13 Comparison of the temperature distribution at the symmetric plane in the solid area of the heat exchanger core under different heat exchange microchannel lengths

Chapter 7
Dissociation Characteristics Simulation of Ammonia, a Propellant for Solar Thermal Propulsion

7.1 Introduction

The ideal working temperature of a solar thermal propulsion (STP) system is generally above 2200 K, and at this temperature, ammonia (NH_3) can dissociate; therefore, it is not accurate to use a single ammonia propellant for the calculation and analysis of thruster performance, and the components of the mixture after ammonia dissociation should be considered. The mixture after ammonia dissociation is composed of atoms and molecules. Since the temperature usually does not exceed 3000 K (it is difficult for high-temperature resistant materials to achieve this temperature), the existence of ionic components can be completely ignored. For vibrational excitation, only molecular nitrogen and hydrogen are considered, which have stable vibrational excitation levels. This chapter focuses on the effect of chemical reactions on the propellant temperature and thruster performance. In an actual 3D flow field, the dissociation characteristics and temperature variation of the ammonia propellant mixture are completely different from those of lumped parameters. The 3D flow field distribution and component distribution pattern are obtained through numerical simulation. This chapter mainly discusses the flow and composition change pattern of the ammonia propellant mixture inside of the heat exchanger core and the nozzle.

7.2 Dissociation Reaction Models and Calculations

7.2.1 Dissociation Reaction Model

The finite-rate chemistry (FRC) model is used for calculations of the ammonia dissociation process. The flow control equations are 3D compressible NS equations and solved by the finite volume method. The discrete format is the second-order upwind scheme.

© National University of Defense Technology Press 2025
M. Huang et al., *Solar Thermal Thruster*, https://doi.org/10.1007/978-981-97-7490-6_7

The main components of ammonia gas after dissociation are N, H, N_2, H_2, NH, NH_2, NNH, and N_2H_2. In the temperature range of 2000–3000 K, the main chemical reactions involved [96] are shown in Table 7.1.

Table 7.1 Arrhenius coefficient of ammonia dissociation reaction

Reaction number	Reaction	A	δ	$E_\alpha/(\text{J·mol}^{-1})$
1	$NH_3 + M = NH_2 + H + M$	2.20×10^{16}	0.00	93,468
2	$NH_3 + H = NH_2 + H_2$	6.36×10^{5}	2.39	10,171
3	$H_2 + M = H + H + M$	2.19×10^{14}	0.00	95,970
4	$NH + M = N + H + M$	2.65×10^{14}	0.00	75,500
5	$NH + H = H_2 + N$	3.60×10^{13}	0.00	325
6	$NH + N = N_2 + H$	3.00×10^{13}	0.00	0.0
7	$NH + NH = N_2 + H + H$	5.10×10^{13}	0.00	0.0
8	$NH_2 + M = NH + H + M$	3.16×10^{23}	−2.0	91,400
9	$NH_2 + H = NH + H_2$	4.00×10^{13}	0.0	3650
10	$NH_2 + N = N_2 + H + H$	7.20×10^{13}	0.0	0
11	$NH_2 + NH = N_2H_2 + H$	1.50×10^{15}	−0.5	0
12	$NH_2 + NH_2 = NH_3 + NH$	5.00×10^{13}	0.0	10,000
13	$NH_2 + NH_2 = N_2H_2 + H_2$	5.00×10^{11}	0.0	0
14	$NNH + M = N_2 + H + M$	2.00×10^{14}	0.0	20,000
15	$NNH + H = N_2 + H_2$	4.00×10^{13}	0.0	3000
16	$NNH + NH = N_2 + NH_2$	5.00×10^{13}	0.0	0
17	$NNH + NH_2 = N_2 + NH_3$	5.00×10^{13}	0.0	0
18	$N_2H_2 + M = NNH + H + M$	5.00×10^{16}	0.0	50,000
19	$N_2H_2 + H = NNH + H_2$	5.00×10^{13}	0.0	1000
20	$N_2H_2 + NH = NNH + NH_2$	1.00×10^{13}	0.0	1000
21	$N_2H_2 + NH_2 = NNH + NH_3$	1.00×10^{13}	0.0	1000

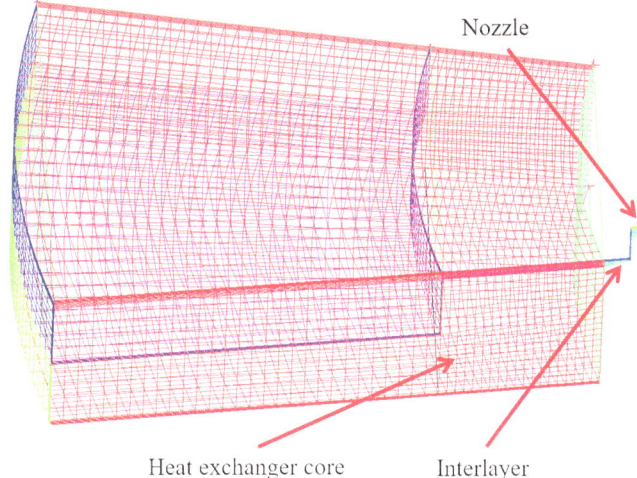

Fig. 7.1 Gridding of the calculation region

7.2.2 Structural Parameters and Gridding

To improve heat exchange efficiency, the laminated microchannel structure is used in the design of the thruster heat exchanger core. Using flow shunting, the heat exchange area between the working fluid and the thrust chamber wall is increased, and the convective heat transfer in the heat exchange channel is improved. Therefore, the working fluid is fully heated in the thrust chamber. In the design, there are 9 laminates, the thickness of a single laminate is 2 mm, and the inner and outer diameters of the laminate are 32 mm and 49 mm, respectively, so the radial length of the laminate is 8.5 mm; the diameter of the control runner is 0.16 mm, and the length of the control runner is 1.5 mm. The calculation region is mainly composed of the following three parts: the fluid–solid coupling heat exchanger core, interlayer and nozzle. The gridding is shown in Fig. 7.1. Due to the axisymmetric characteristics of the thruster structure, to reduce the calculation volume, half of the single runners in the laminate structure are taken in the calculation. The upper, lower, left, and right sides of the model are symmetric planes, and the circumferential angle in the model is 22.5°.

7.2.3 Boundary Conditions

The working fluid inlet pressure $P_c = 0.8$ MPa, and the inlet temperature $T_i = 300$ K. In the calculations considering the ammonia dissociation reaction, the component at the inlet is NH_3 only. The nozzle exhaust is under vacuum, the pressure $P_e = 0$, and the exhaust temperature $T_e = 300$K. The solar radiation power on the inner wall of the heat exchanger core is 1.2×10^6 W/m^2, and the outer wall is set to adiabatic conditions.

7.3 Effect of Dissociation Reaction of Ammonia Propellant on the Heating Effect of the Heat Exchanger Core

7.3.1 Simulation of 1D Flow Characteristics

To understand the overall high-temperature dissociation characteristics of ammonia, the CHEMKIN program is used to calculate the dissociation reaction of ammonia in the flow process, and the parametric characteristics of the 1D flow of ammonia high-temperature dissociation are obtained. The runner length is 0.3 m. In the calculations, the initial temperature of ammonia is given, and the variation pattern of the ammonia mixture components along the flow direction is calculated. Figure 7.2 shows the dissociation characteristics of the ammonia propellant at 2400 K. H_2 and N_2 are the main products, and the concentrations of intermediates N, H, NH, NH$_2$, NNH and N$_2$H$_2$ are very small and close to 0 in the figure.

Figure 7.3 shows the variation pattern of the average molar mass of the dissociation mixture at three temperatures of 2200, 2400 and 2600 K. As the temperature increases, the ammonia dissociation increases, and the average molar mass of the mixture decreases. Along the flow path of the working fluid, the average molar mass of the working fluid also decreases significantly. Taking 2400 K as an example, when equilibrium is reached, the average molar mass is already less than 11 g/mol.

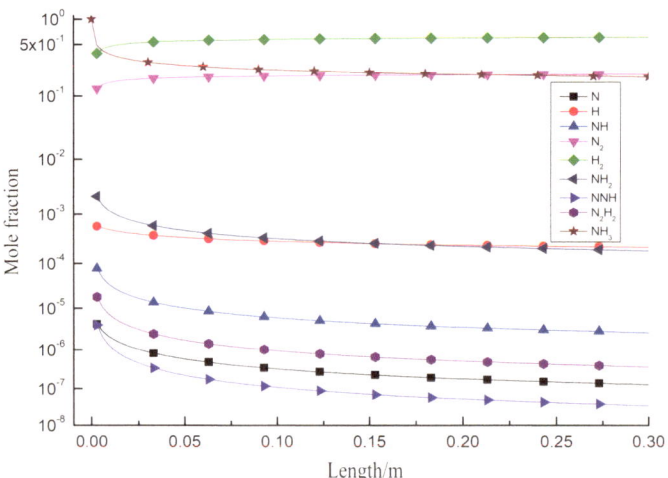

Fig. 7.2 Changes in the dissociation components of ammonia propellant at 2400 K

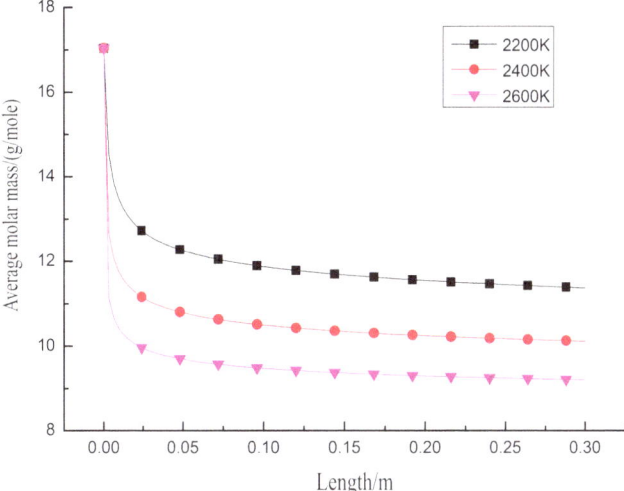

Fig. 7.3 Average molar mass change of the mixture

Figure 7.4 shows the effect of temperature on the mole fraction of the ammonia propellant after being heated, and Fig. 7.5 shows the effect of heating temperature on the dissociation of ammonia gas. As shown in Fig. 7.4, as the temperature increases, the dissociation of ammonia intensifies. When the temperature exceeds 2600 K, the mole fraction of ammonia in the mixture at equilibrium is only 0.1, and the degree of dissociation is 0.85. On the other hand, when the temperature is 2200 K, the mole fraction of the mixture is approximately 0.4, and the degree of dissociation is 0.50. The degree of dissociation is calculated based on the temperature consistency in the ammonia propellant flow. In the actual flow field, due to the nonuniform distribution of temperature and speed, the actual degree of dissociation is lower than the calculated value, and the mole fraction of ammonia is larger.

7.3.2 Simulation of Distributed Parameter Characteristics

The initial state of the propellant at the inlet is ammonia gas. Under this working condition, a 3D fluid–solid coupling heat transfer model is established to simulate the dissociation reaction.

Figure 7.6 shows the 3D distribution cloud map of the temperature of the ammonia propellant mixture, and Fig. 7.6a and b show different angles. In the laminated heat exchanger core, since the ammonia dissociation process is an endothermic reaction, the calculated temperature distribution when the dissociation reaction is considered is slightly lower than that without considering the dissociation reaction. The temperature distribution of the heat exchanger core sections is shown in Fig. 7.7, with the left showing the temperature distribution without considering the dissociation reaction

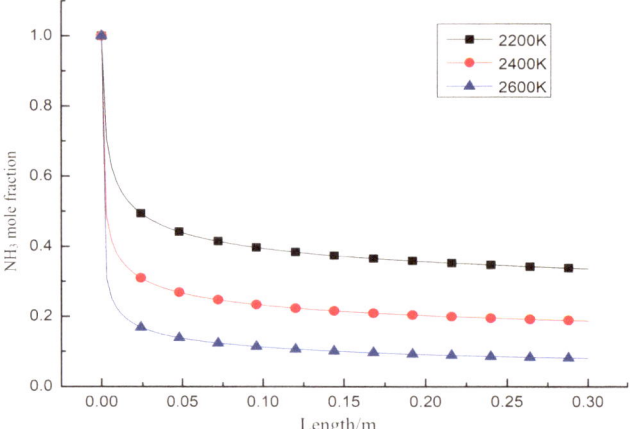

Fig. 7.4 Changes in the mole fraction of ammonia at different temperatures

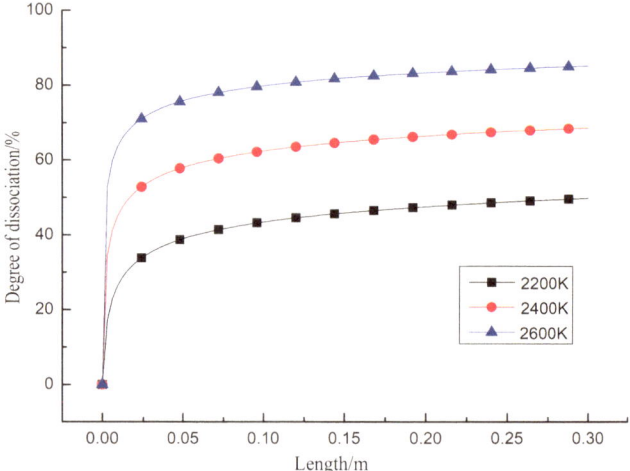

Fig. 7.5 Changes in the degree of dissociation of ammonia at different temperatures

and the right showing the temperature distribution when considering the dissociation reaction. Figure 7.7 shows that, for the laminated heat exchanger core, because of the fast flow speed of the working fluid in the control runner, the temperature of the dispersion area close to the control runner is relatively low, and the temperature in other locations in the dispersion area is high. However, after further heating of the interlayer, the flows with temperature differences meet in the interlayer, and then the temperature gradually becomes the same. When reaching the front end of the nozzle, the temperature difference is very small. The average temperature at the nozzle inlet

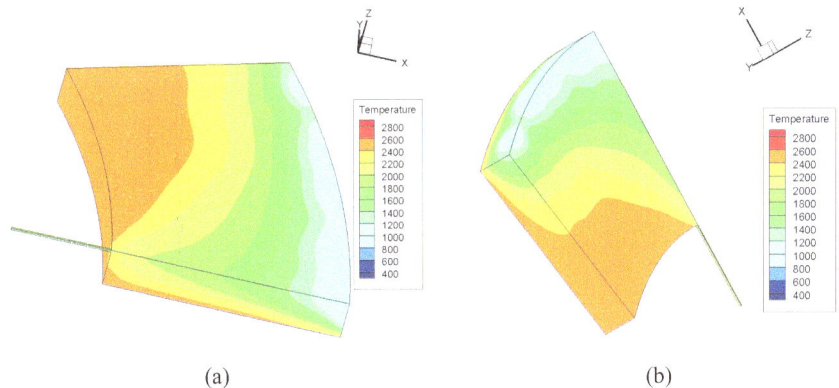

Fig. 7.6   3D distribution cloud map of the temperature of ammonia propellant mixture gas

when considering the dissociation reaction is 2323 K, which is very close to the calculation result without considering the dissociation reaction (2340 K).

Figure 7.8 shows the changes in the mass fraction distribution of the ammonia dissociation components in the heat exchanger core.

As shown in Fig. 7.8, the dissociation of ammonia in the control runner is not sufficient, and the mass fraction of ammonia in the mixture is still more than 30%. With diffusion in the dispersion area and increasing temperature, the dissociation of ammonia continuously increases.

After the heat exchanger core and the interlayer are heated, the mass fraction and mole fraction distributions of each component after the ammonia gas is dissociated before entering the nozzle are shown in Table 7.2. The average molar mass at the nozzle inlet is 10.036 g/mol. The dissociation reaction process during the heating process is not sufficient, and the dissociation reaction of ammonia gas after the heat exchanger core and the interlayer are heated is insufficient, with the degree of dissociation being 0.143.

7.4   Effect of Dissociation Reaction of Ammonia Propellant on Nozzle Performance

In this section, the effect of the dissociation reaction of ammonia propellant on the performance of the nozzle is analyzed. The flow speed vector distribution in the nozzle without considering the ammonia dissociation reaction is shown in Fig. 7.9. After integration, the specific impulse of the nozzle is 219 s. Figure 7.10 shows the flow speed vector distribution in the nozzle considering the ammonia dissociation reaction. The specific impulse of the system can increase to 251 s with the consideration of the chemical reaction. The main reason is that after the dissociation of ammonia, the

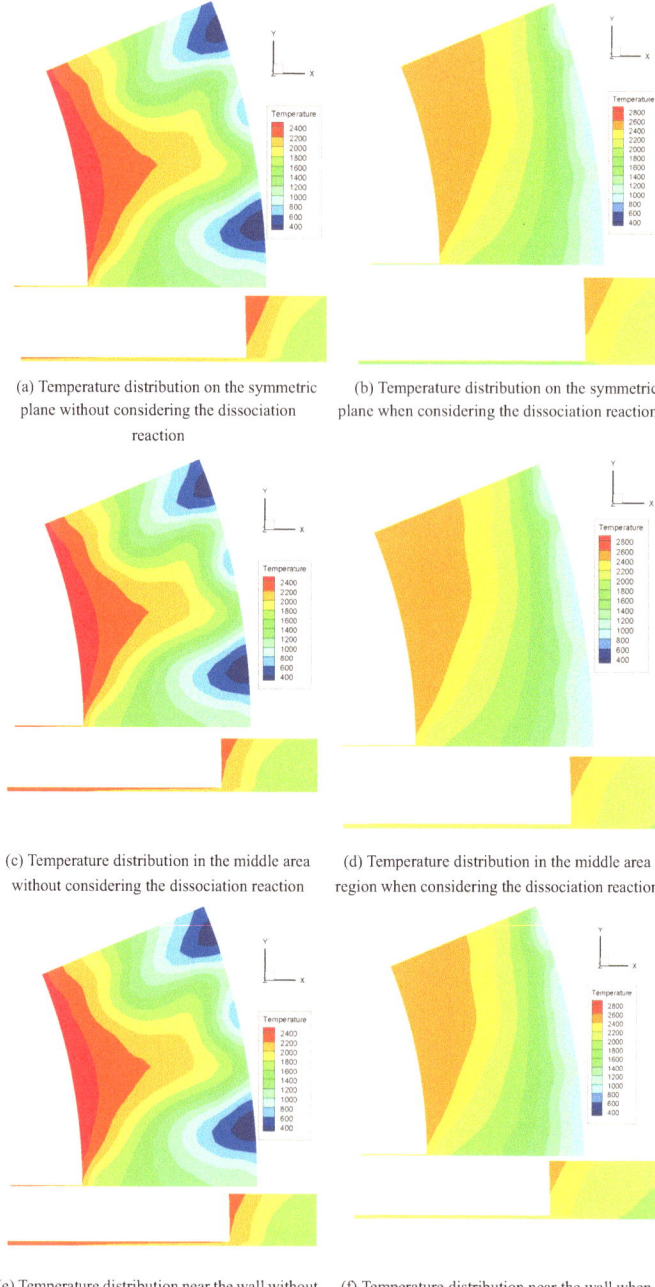

(a) Temperature distribution on the symmetric plane without considering the dissociation reaction

(b) Temperature distribution on the symmetric plane when considering the dissociation reaction

(c) Temperature distribution in the middle area without considering the dissociation reaction

(d) Temperature distribution in the middle area region when considering the dissociation reaction

(e) Temperature distribution near the wall without considering the dissociation reaction

(f) Temperature distribution near the wall when considering the dissociation reaction

Fig. 7.7 Comparison of the temperature of the heat exchanger core sections

Fig. 7.8 Changes in the distribution of the ammonia dissociation components in the heat exchanger core

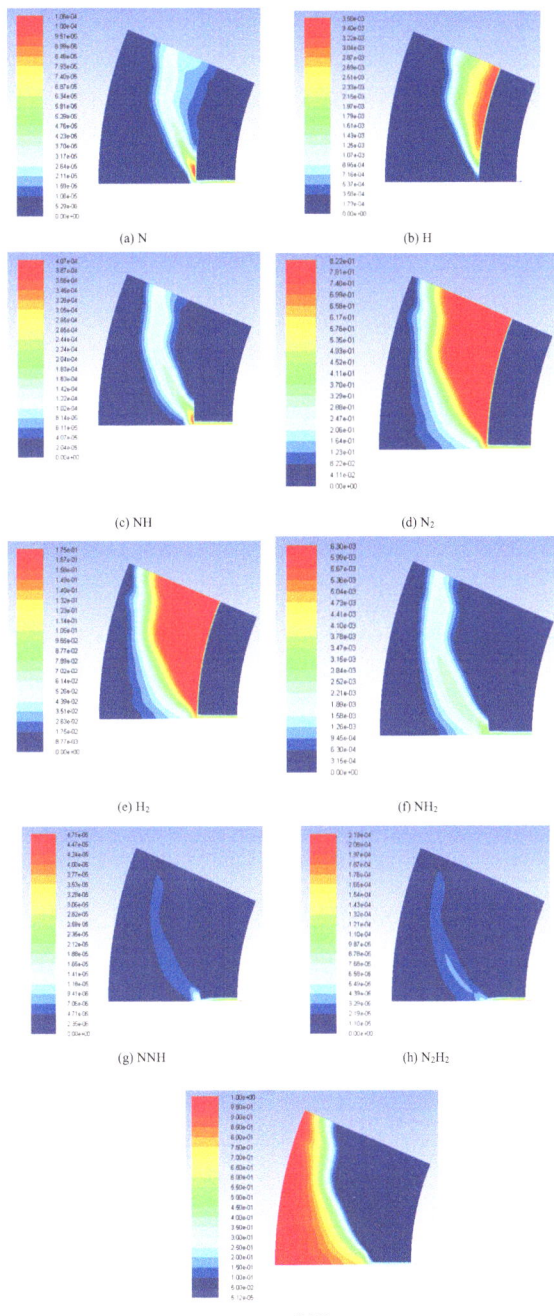

Table 7.2 Component distribution at the nozzle inlet

Component	Mass fraction	Mole fraction
N	7.4×10^{-5}	5.31×10^{-5}
H	2.37×10^{-4}	2.36×10^{-3}
NH	2.98×10^{-4}	1.99×10^{-4}
N_2	0.568	0.204
H_2	0.123	0.612
NH_2	3.01×10^{-3}	1.88×10^{-3}
NNH	3.32×10^{-5}	1.15×10^{-5}
N_2H_2	1.64×10^{-4}	5.47×10^{-5}
NH_3	0.305	0.180

average molar mass of the working fluid decreases. The calculated average molar mass of the working fluid along the axial direction of the nozzle decreases from 14.8 to 13.8 g/mol. After passing through the nozzle throat, the dissociation reaction no longer occurs due to the rapid decrease in the temperature of the working fluid, and the average molar mass is also no longer changed. On the other hand, the molar mass of the ammonia propellant without considering the chemical reaction is 17 g/mol, and the change in the specific impulse is consistent with the theoretical increase in the specific impulse brought about by the decrease in molar mass. With the consideration of the ammonia dissociation reaction, the actual performance of the thruster is greatly improved, and ammonia gas is more competitive as a propellant.

If the temperature of the thruster heat exchanger core is further increased, the dissociation of ammonia can increase, and the specific impulse of the thruster can be further increased. When the dissociation reaction is not considered, the nozzle inlet

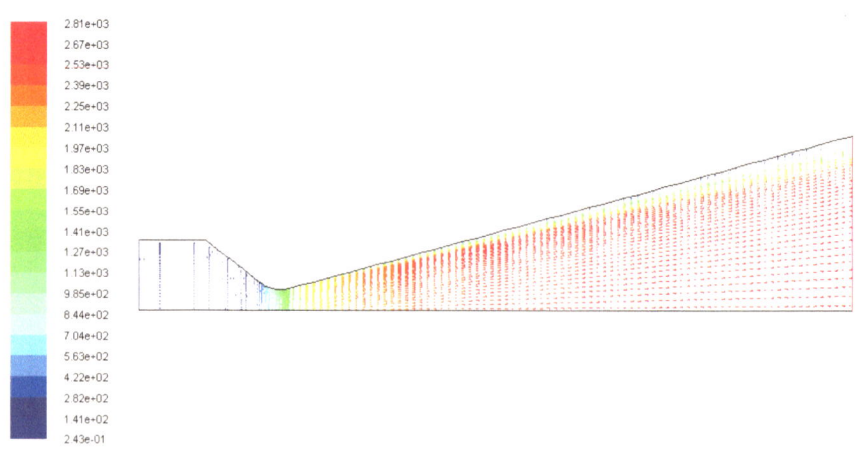

Fig. 7.9 Flow speed vector distribution in the nozzle without considering the ammonia dissociation reaction (unit: m/s)

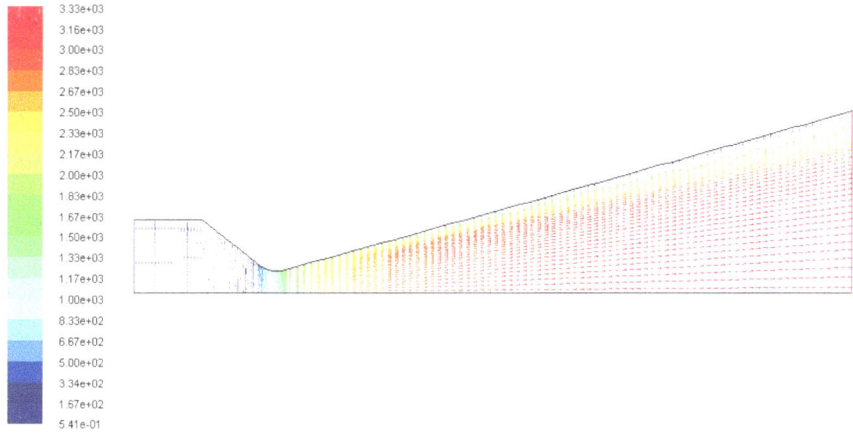

Fig. 7.10 Flow speed vector distribution in the nozzle considering the ammonia dissociation reaction (unit: m/s)

temperature is 2540 K, and the specific impulse of the thruster is 224 s. When the dissociation reaction is considered and the nozzle inlet temperature is 2540 K, the degree of dissociation of ammonia is 43%, and the specific impulse of the thruster can reach 302 s, indicating that the performance of the thruster is greatly improved. Therefore, although the ammonia dissociation reaction can slightly lower the temperature of the working fluid when mixed with components compared to the case of a single working fluid, the increase in the specific impulse of the thruster after passing the nozzle is not affected.

Figure 7.11 shows a mass distribution cloud map of each component in the nozzle under the condition that the heat exchanger core wall temperature is 2400 K. Figure 7.11 shows that due to the high temperature in the nozzle contraction section, the dissociation reaction still proceeds, and therefore, the mass fraction of NH_3 still decreases continuously. In the nozzle expansion section, due to the low temperature, the reaction basically no longer occurs; therefore, the mass fraction of the NH_3 component basically remains unchanged, and it is difficult for other components to recombine into NH_3. The distribution patterns of the N_2 and H_2 components are similar. The concentration in the nozzle contraction section increases, and after passing through the throat, the concentration in the diffusion section basically remains unchanged, and the distribution is uniform. The concentrations of other components are all low, and these components are intermediate products with a short residence time before being quickly converted to more stable N_2 and H_2. For components with low concentration, such as N, H, and N_2H_2, the changing pattern is similar. The concentration increases along the nozzle flow direction, and the distribution characteristics are similar to the temperature cloud map, indicating that as the temperature decreases, the formation reactions of N, H, and N_2H_2 dominate, and the concentrations increase. However, since the overall concentration is only on the order of 10^{-5}–10^{-4}, the impact on nozzle performance is very slight. The changes in NH and

NH_2 concentrations are similar, with the concentrations decreasing along the flow direction of the nozzle. The concentrations are on the order of 10^{-4} and 10^{-3}. The concentration of NNH decreases in the contraction section, has a minimum value in the throat, and then gradually increases in the expansion section, with a concentration on the order of 10^{-5}. Analysis shows that the molar factions of N, H, NH, NH_2, NNH and N_2H_2 in the final products are very small, which does not affect the performance of the thruster nozzle, but they are important intermediate products to generate the final products N_2 and H_2, and their role cannot be ignored. The change in the mole fraction of each component in the nozzle when the heat exchanger core wall temperature is 2600 K is shown in Fig. 7.12.

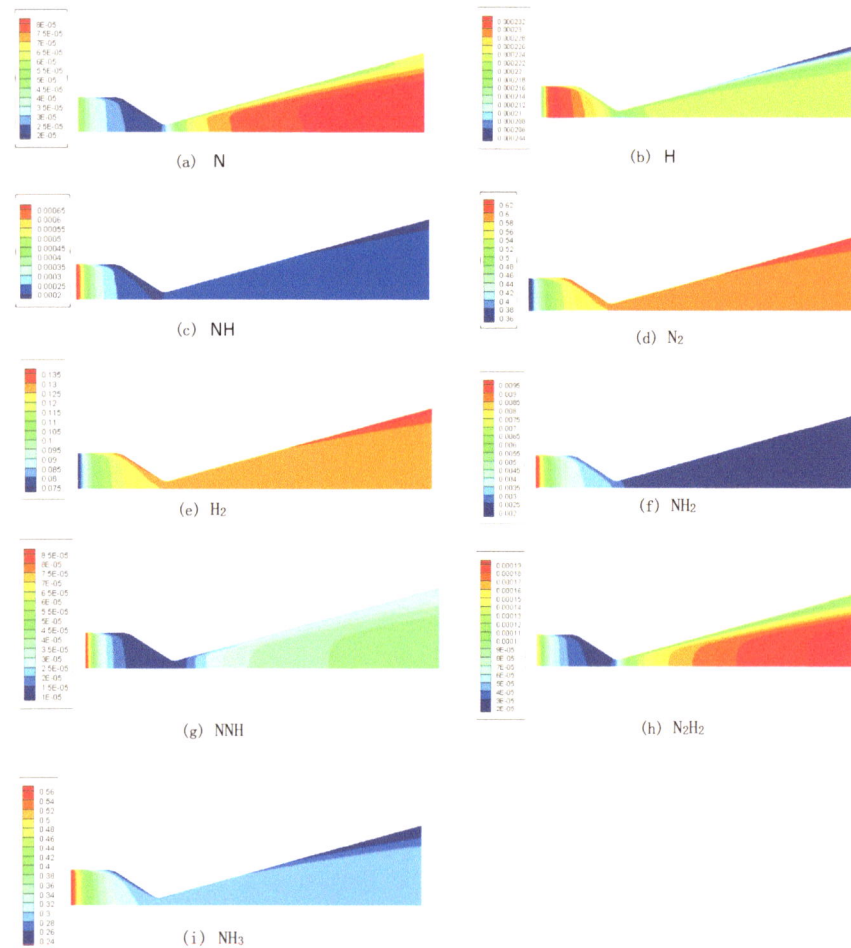

(a) N

(b) H

(c) NH

(d) N_2

(e) H_2

(f) NH_2

(g) NNH

(h) N_2H_2

(i) NH_3

Fig. 7.11 Changes in the mass fraction distribution of each component in the nozzle

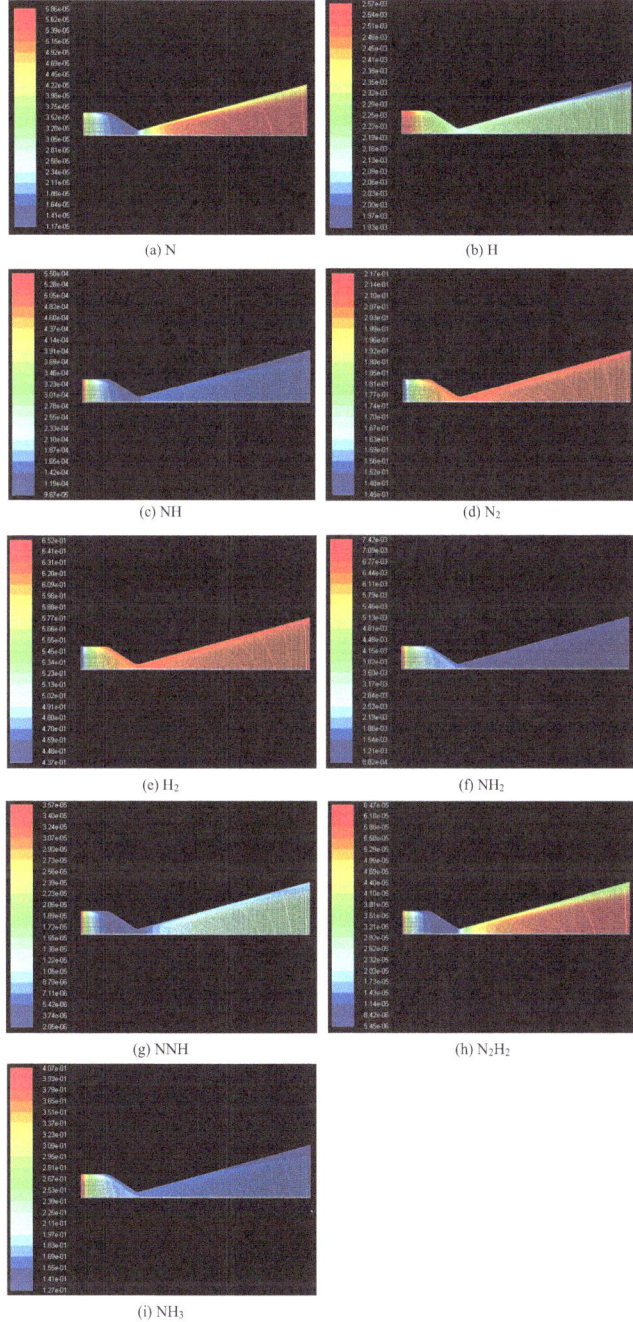

Fig. 7.12 Changes in the mole fraction distribution of each component in the nozzle

The changes in the mole fractions of the main dissociation components, N_2, H_2 and NH_3, along the nozzle axis are shown in Fig. 7.13. The mole fractions of N_2 and H_2 first increase at the nozzle inlet and then do not change. The reason is that the temperature inside the nozzle is not high enough to maintain the dissociation reaction, and there is no change in the nozzle expansion section, which is in a state of frozen flow.

Figure 7.14 shows the variation pattern of all components. Figure 7.14 shows that the variation characteristics of the dissociation products in the nozzle are similar to those of the 1D flow simulation in the previous section. The mole fractions of intermediates such as N and H in the nozzle are very small because intermediates such as N and H are unstable and can further react to form N_2 and H_2.

Figure 7.15 shows the trend of each elementary reaction rate in the nozzle. Figure 7.15 shows that the reactions are mainly concentrated at the nozzle inlet. As the temperature of the produced mixture in the nozzle decreases, the reaction rate gradually decreases to 0, and no elementary reactions occur.

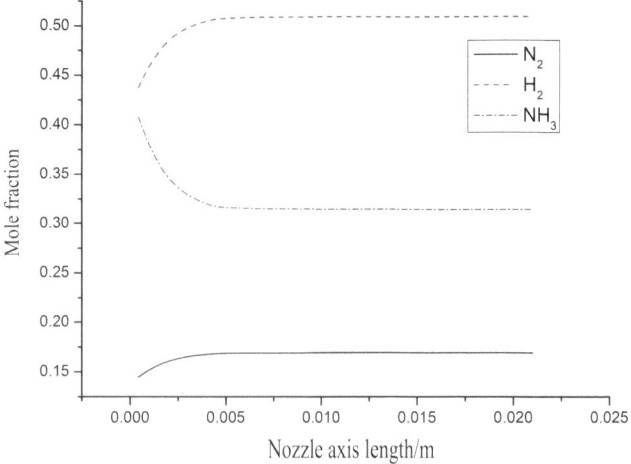

Fig. 7.13 Changes in N_2, H_2 and NH_3 along the nozzle

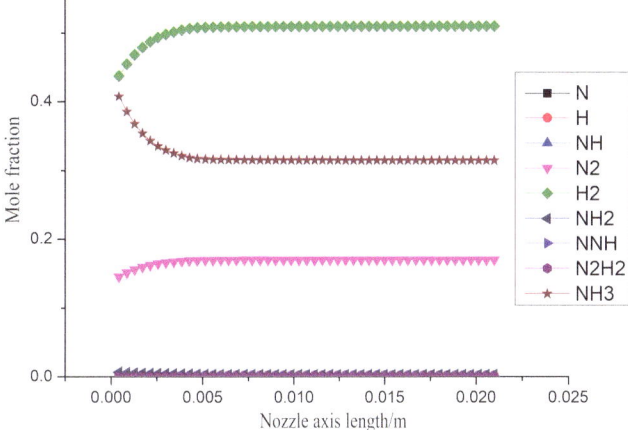

Fig. 7.14   Mole change of dissociation products along the nozzle

Fig. 7.15   Trend of each elementary reaction rate in the nozzle

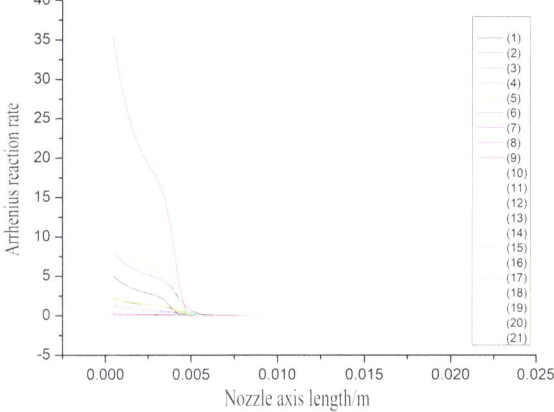

Chapter 8
Integrated Design of Solar Thermal Propulsion and Task Optimization

8.1 Introduction

In previous chapters, the secondary concentrator, absorption cavity, heat exchanger core, and nozzle of the solar thermal propulsion (STP) system are integratedly designed, and regenerative cooling and laminated heat exchange technologies are used to effectively combine these components. This chapter analyzes the overall efficiency of the STP system and uses quantum genetic algorithms to perform a multiobjective optimization for space missions.

8.2 Efficiency Analysis of STP System

Taking the propellant of the solar thermal thruster as the study object, according to the energy balance relationship, the changing rate of the enthalpy rise of the propellant gas is equal to the difference between the solar power received by the thruster and the heat loss rate of the thruster to the outside:

$$\dot{m}(h_i - h_e) = \dot{Q}_{\text{solar}} - \dot{Q}_{\text{loss}} = \dot{Q}_{\text{solar}} - (1 - \eta)\dot{Q}_{\text{solar}} = \eta\dot{Q}_{\text{solar}} \tag{8.1}$$

where η is the heat transfer efficiency of the solar thermal thruster, h_i, h_e are the specific enthalpy of the propellant at the inlet and outlet of the heat exchanger, $\dot{Q}_{\text{solar}}$ is the solar power received by the primary concentrator, and $\dot{m}$ is the mass flow rate of the propellant.

The thermal efficiency of STP system is defined as:

$$\eta_t = \frac{\dot{m}h_e + \frac{1}{2}\dot{m}v_e^2}{\dot{m}h_i + P} \tag{8.2}$$

© National University of Defense Technology Press 2025
M. Huang et al., *Solar Thermal Thruster*, https://doi.org/10.1007/978-981-97-7490-6_8

Propulsion efficiency can be defined as

$$\eta = \frac{\dot{m}v_e^2}{2(\dot{m}h_i + P)} \tag{8.3}$$

In the design of the thrust chamber parameters, the temperature of propellant hydrogen (1300 K) is selected as the qualitative temperature, and other parameters are as follows: the density $\rho = 0.0189$ kg/m^3, the molar mass $M = 2.016 \times 10^{-3}$ kg/mol, the specific heat ratio $\gamma = 1.404$, the average specific heat capacity at constant pressure $C_p = 1.56 \times 10^4$ J/(kg K), the dynamic viscosity $\mu = 24.08 \times 10^{-6}$ kg/(m s), and the average thermal conductivity $k = 0.568$ W/(m K).

First, the effects of thrust chamber pressure and concentration ratio on thruster performance are analyzed. Based on the simulation results of the laminated heat exchanger core and regenerative cooling in the previous chapters, the performance parameters of the thruster, such as the specific impulse, thrust and thermal efficiency, are calculated.

The variation in the total propellant temperature versus the concentration ratio under different thrust chamber pressures is shown in Fig. 8.1. As the concentration ratio increases, the possible total propellant temperature gradually increases, and with the increase in thrust chamber pressure, the total propellant temperature shows an overall decreasing trend. At pressures greater than 0.4 MPa, it is difficult for the total propellant temperature to reach 2000 K. The variation in the thermal efficiency of the system versus the concentration ratio under different thrust chamber pressures is shown in Fig. 8.2. As shown in Fig. 8.2, under the condition of a constant thrust chamber pressure, the thermal efficiency of the thruster shows an upward trend as the concentration ratio increases, and with the increase in the thrust chamber pressure, the thermal efficiency of the thruster shows an overall increasing trend. The trend of thermal efficiency to increase with the increase in the concentration ratio also gradually decreases. When the chamber pressure is greater than 0.3 MPa, the effect of the change in the concentration ratio on the thermal efficiency is negligible. The chamber pressure of 0.3 MPa is the critical point under this working condition, and the variation in thermal efficiency is ignored when the concentration ratio is greater than 5000. On the other hand, when the chamber pressure is 0.2 MPa, the efficiency is low.

For the propulsion efficiency of the system, when the chamber pressure is greater than 0.3 MPa, the effect of the concentration ratio on the efficiency is negligible, as shown in Fig. 8.3. When the concentration ratio is greater than 5000, the efficiency at the chamber pressure of 0.3 MPa is very close to that at 0.4–0.8 MPa. On the other hand, when the chamber pressure is 0.2 MPa, the efficiency is low.

Under different pressures, the variations in the specific impulse and thrust versus the concentration ratio are shown in Figs. 8.4 and 8.5. As shown in Fig. 8.4, as the concentration ratio increases, the specific impulse obtained by the thruster gradually increases, and as the pressure increases, the specific impulse of the thruster shows an overall decreasing trend. As shown in Fig. 8.5, as the concentration ratio increases, the thrust obtained by the thruster shows a decreasing trend. In addition, as the

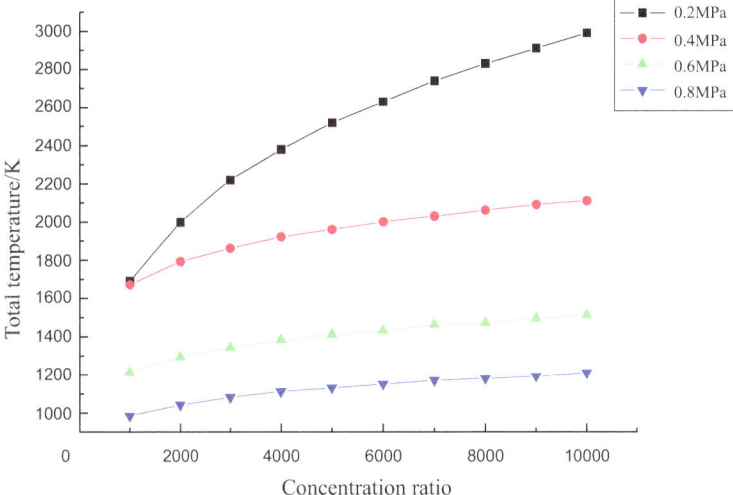

Fig. 8.1 Variation in total temperature versus concentration ratio under different thrust chamber pressures

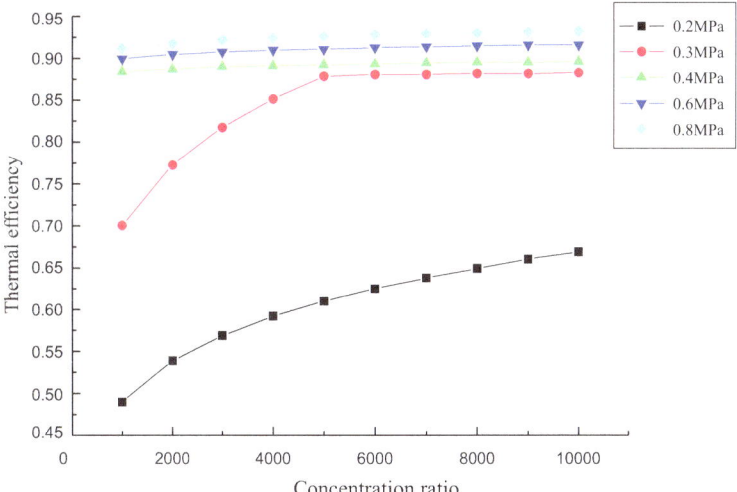

Fig. 8.2 Variation of thermal efficiency of the system versus concentration ratio under different thrust chamber pressures

pressure increases, the thrust of the thruster shows an overall rising trend, while when the pressure is low, the variation in the thrust versus the concentration ratio is not significant.

According to the analysis results, when the chamber pressure is high, such as 0.8 MPa, greater thrust and higher thermal efficiency can be obtained, but the achievable

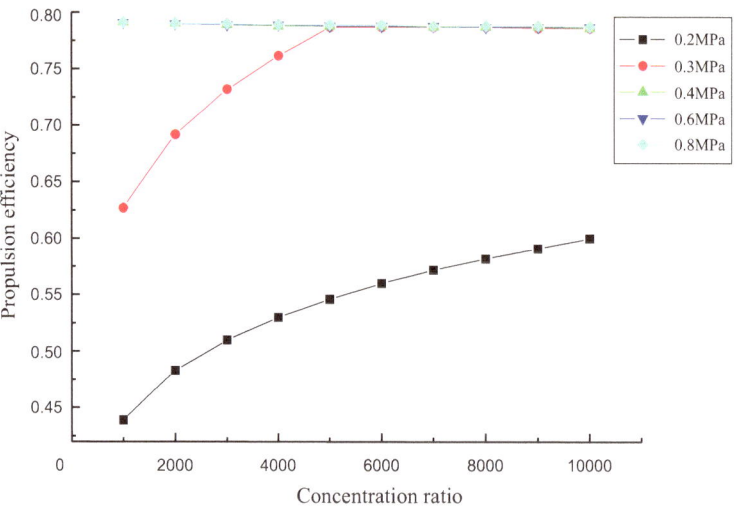

Fig. 8.3 Variation in propulsion efficiency of the system versus concentration ratio under different thrust chamber pressures

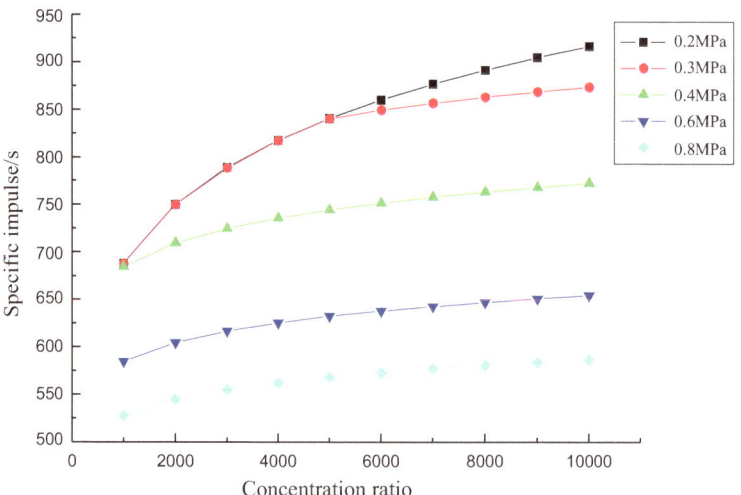

Fig. 8.4 Variation in the specific impulse versus concentration ratio under different thrust chamber pressures

kinetic energy efficiency of the system does not increase significantly compared to that at 0.3 MPa. Under this condition, the specific impulse is too low, so it is not an ideal choice. A comprehensive comparison shows that a chamber pressure of 0.3 MPa is a more suitable choice, the specific impulse can be greater than 800 s, and the thrust should be approximately 2 N. Similarly, for concentrators with other

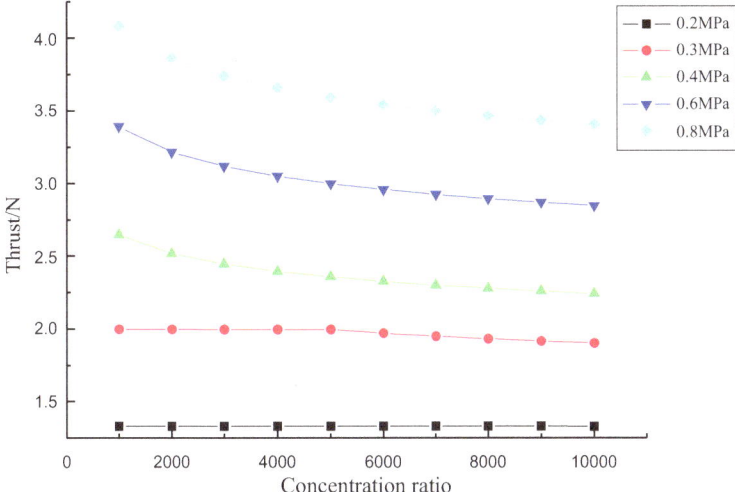

Fig. 8.5 Variation in thrust versus concentration ratio under different thrust chamber pressures

diameters, there exists a superior choice of chamber pressure. It is feasible to set the chamber pressure according to the specific impulse and thrust needed for different space tasks.

The specific impulse is an important performance parameter in a space propulsion system. Under the premise of ensuring the maximum specific impulse of the system, the influences of the concentrator diameter and the concentration ratio on the specific impulse, thrust, and thermal efficiency of the system are analyzed.

Under a certain concentration ratio, the highest temperature that the absorption cavity can obtain is fixed, and the corresponding specific impulse is the highest one that can be obtained under this concentration ratio, regardless of the influence of the input energy. The variation pattern of the maximum temperature and maximum specific impulse of the absorber of thruster versus the concentration ratio is shown in Fig. 8.6. As the concentration ratio increases from 1000 to 10000, the highest temperature of the absorber increases from 1680 to 2990 K, and the maximum specific impulse increases from 687 to 916 s.

The variation in the thermal efficiency of the system versus the concentration ratio under different diameters of the primary concentrator is shown in Fig. 8.7. As shown in Fig. 8.7, as the concentration ratio increases, the thermal efficiency of the thruster decreases, and with the increase in concentrator diameter, that is, the increase in incident solar power, the overall thermal efficiency shows an overall increasing trend. The decreasing trend of thermal efficiency also gradually decreases as the concentration ratio increases. For a propulsion system using 1 m diameter concentrators, the thermal efficiency of the thruster is most affected by the change in the concentration ratio. With the increase in the concentration ratio, the thermal efficiency decreases from 0.845 to 0.796, with a large variation range. On the other

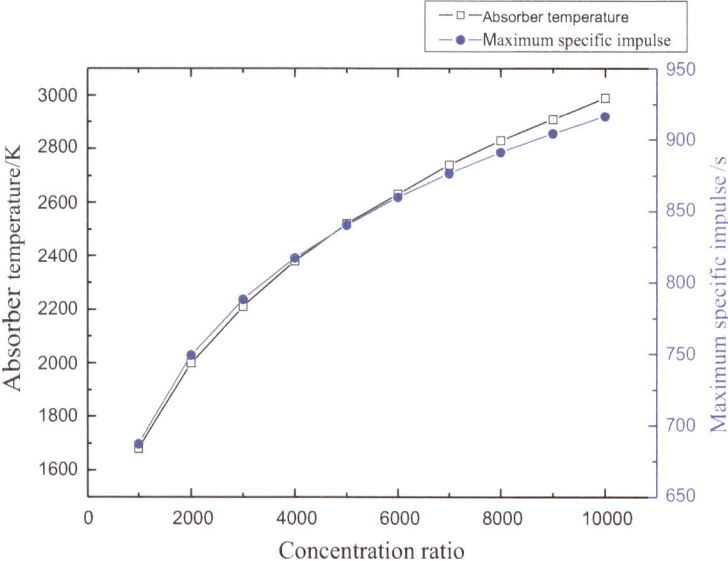

Fig. 8.6 Variations in the maximum temperature and specific impulse of the thruster versus the concentration ratio

hand, for the propulsion system using 4 m diameter concentrators, the thermal efficiency of the thruster is least affected by the change in concentration ratio; under the same change in concentration ratio, the thermal efficiency only decreases from 0.884 to 0.881. Under the same concentration ratio, the temperature of the thruster is a fixed number, and the corresponding heat loss is also a fixed number. Therefore, when the input solar power is small, the thermal efficiency is correspondingly low. Heat loss accounts for a large proportion of the total energy, and thus the impact on thermal efficiency is severe.

The variation in the thrust versus the concentration ratio under the different primary concentrator diameters is shown in Fig. 8.8. As shown in Fig. 8.8, as the concentration ratio increases, the thrust obtained by the thruster shows a decreasing trend, and with the increase in concentrator diameter, that is, the increase in incident solar power, the thrust of the thruster shows an overall increasing trend, which is the same as the change in thermal efficiency. However, different from the change in thermal efficiency, the decreasing trend of the thrust with the increase in the concentration ratio gradually increases with the increase in diameter. For the propulsion system using 4 m diameter concentrators, the thrust is most affected by the change in concentration ratio; with the increase in concentration ratio, the thrust decreases from 4.682 to 3.195 N, with a large variation range. On the other hand, for the propulsion system using 1 m diameter concentrators, the thrust of the thruster is least affected by the change in concentration ratio. Under the same change in concentration ratio, the thrust only decreases from 0.277 to 0.178 N.

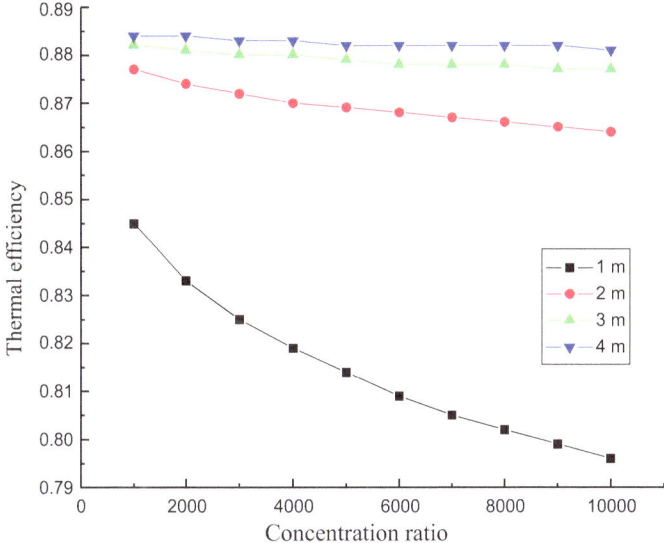

Fig. 8.7 Variation of thermal efficiency of the system versus concentration ratio under different primary concentrator diameters

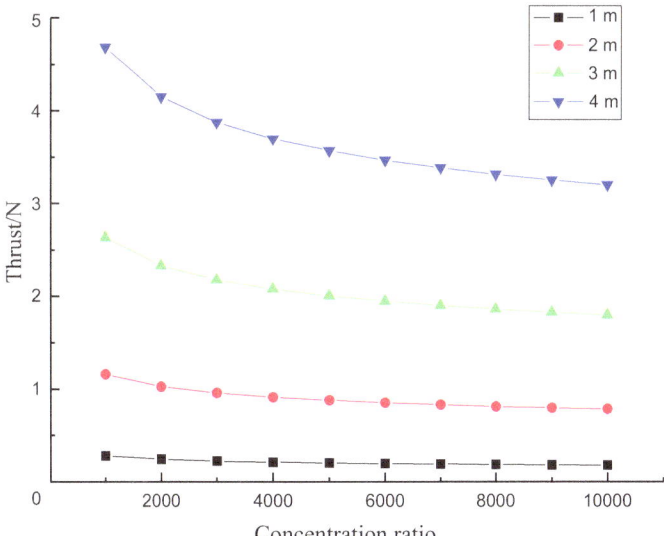

Fig. 8.8 Variation in thrust versus condensation ratio under different primary concentrator diameters

8.3   Application and Optimization of Solar Thermal Propulsion

8.3.1   Applications for Space Propulsion

The power system used for space propulsion requires good performance, high reliability, simple structure, long service life, and the ability to be started in space multiple times. STP technology can meet these performance requirements. The space propulsion system and the payload are sent into a predetermined orbit by the launch vehicle, and the propellant used in the system must be carried during the process from takeoff to the predetermined orbit, so the structure mass of the space propulsion system affects the payload, and the propellant mass carried also affects the lifetime of the propulsion system and the spacecraft. In other words, increasing the specific impulse of the propulsion system or greatly reducing the weight of the space propulsion system can extend the working time of the spacecraft in orbit, thereby increasing the service life of the spacecraft and reducing the cost. Monopropellant and double-propellant chemical rocket propulsion systems are highly developed, and this mature technology and durable can provide spacecraft with the long-term storage and intermittent start-up capability needed for various flight maneuvers. However, the specific impulse of the chemical propulsion system is relatively low (the maximum specific impulse of the double-propellant chemical propulsion system is approximately 450 s), and the system is complex and bulky. If chemical propulsion systems are used for special tasks, such as orbital transfer, orbital correction, attitude control, docking and rendezvous, position maintenance, north–south station-keeping (NSSK) orbit control and interplanetary navigation of space vehicles, the launching costs of space vehicles, which have always been expensive, will remain high, and the service life could be seriously affected, thus making long-distance interplanetary voyages and deep-space exploration impractical.

At present, all the major leading countries in space exploration are looking for novel small-thrust nonchemical propulsion systems that offer high performance, low cost, small volume, light weight, and low power consumption. Among them, electric propulsion is attractive and technically easy to achieve. Electric propulsion systems have the characteristics of high specific impulses, small thrusts, light weights, small volumes, low working fluid consumption, and certain electric power requirements. The use of electric propulsion with a high specific impulse (up to 300–5000 s) can save fuel and increase payload, which is important for deep space exploration, especially interplanetary exploration. The only disadvantage is that the thrust of an electric propulsion system is very slight (0.001–1 N), and it takes a long time to complete the acceleration process, which adds difficulty to the design of orbit transfer, ascension, deep space exploration, and interstellar navigation missions when using small thrust propulsion systems. For example, SMART-1, the first European Space Agency satellite that orbited the Moon, used an ion thruster engine as the electric propulsion system. After the satellite was launched into space, it took 13 and a half months to reach the Moon. STP systems, due to their relatively high specific impulse (up to 800 s with hydrogen as the propellant) and moderate thrust (0.4–100 N), have attracted the attention of researchers in the aerospace field. The performance of an STP system ranges between chemical propulsion and electric propulsion systems, filling the performance gap of chemical propulsion and electric propulsion systems. As the primary and auxiliary propulsion system of a spacecraft, STP can reduce the launch cost and increase the payload. STP systems are particularly suitable for missions that do not require high thrust. The Marshall Space Flight Center (MSFC) studied a space transfer vehicle that uses STP, which can transfer a payload of 450 kg from a low Earth orbit (LEO) to a geostationary Earth orbit (GEO). In the design, liquid hydrogen is used as a propellant, and two inflatable off-axis parabolic concentrators are used to guide the concentrated sunlight into a tungsten/rhenium blackbody absorber. The concentration ratio of the primary concentrator is 1800:1, the concentration ratio of the secondary concentrator is 3:1, and the total concentration ratio is 5400:1. The generated thrust is 8.9 N, and the specific impulse is 860 s. The transfer time to GEO is approximately one month.

A performance comparison of the solar thermal propulsion system and other propulsion systems is shown in Table 8.1. The comparison shows that the STP system has the advantages of high specific impulses and the ability to generate appropriate thrusts, and these advantages determine its wide application in the fields of satellite orbital transfer, interstellar navigation missions, and powering microsatellites. STP can only provide a small thrust, and is characterized by a high specific impulse, sustainable propulsion, and adjustable and accurate thrust. Compared with chemical propulsion, low-thrust propulsion can better meet the demand for frequent satellite orbit transfers and has a longer orbit lifetime.

Table 8.1 Performance comparison of propulsion systems

Classification of propulsion methods			Typical thrust/N	Typical specific impulse/s
Electric propulsion	Electric heating	Electric resistance heating	0.05	180
	Electrostatic	Ion thruster	5×10^{-3}	3000
		Colloid thruster	$10^{-6}–10^{-5}$	500
	Electromagnetic	Hall-effect thruster (HET)	$10^{-3}–10^{-1}$	1600
Chemical propulsion	*Solid propulsion*		0.1	100–300
	Liquid propulsion	Monopropellant propulsion	0.02–0.75	150
		Double-propellant propulsion	15	290
Cold gas propulsion			0.5–4.5	60
Solar thermal propulsion			0.02–2	200–900

At present, the space missions suitable for STP can be mainly divided into three categories:

(1) Near-escape mission: The ideal speed increment from a geosynchronous transfer orbit (GTO) to a final orbit is 700–1000 m/s, and a short period and ignition at the perigee point are needed to reach the final orbit.
(2) GEO mission: The requirement (GTO-GEO) is on the order of 1500 m/s, allowing a small thrust at the apogee point to accelerate to reach the final orbit.
(3) The capture missions of other celestial bodies include the lunar orbit, the GTO to the final orbit with an ideal speed increment in a range of 1100–1500 m/s, and interplanetary missions. These missions all require low-thrust systems to provide the combined ignitions at perigee and apogee points.

An STP system can provide a continuous thrust of 1–5 N or an intermittent thrust of 10–100 N, and the continuous constant thrust is a commonly used orbital maneuver in space flight. Among them, small thrusts are suitable for the rendezvous maneuvers of Earth-orbiting spacecraft, while tangential or circumferential thrusts and larger positive radial thrusts can be used to escape flight from the Earth's gravitational field and perform interstellar object rendezvous. Using the motion equation of the center of mass under a constant thrust, there is no limitation on the magnitude of the maneuvering thrust, and during spacecraft rendezvous applications, there is no requirement on the relative distance.

8.3.2 Optimization of Computational Model

For spacecraft needing space propulsion, small satellites are the main design objects, and there are two types. The total mass of the first type of satellite is 100 kg, the volume is $60 \times 60 \times 80$ cm (0.288 m^3), the volume of the propellant tank is 0.05 m^3, and the mass is limited to 50 kg.

The total mass of the second type of satellite is 400 kg, the volume is $110 \times 110 \times 88.5$ cm (1.07 m^3), the volume of the propellant tank is 0.18 m^3, and the mass is limited to 200 kg.

8.3.2.1 Objective Function

(1) Speed increment ΔV

Under the thrust of a solar thermal thruster, the best orbital transfer effect can be achieved by maximizing the speed increment ΔV. Achieving the best match between the spacecraft and thruster performance is the most direct reflection of the maximation of the speed increment ΔV, so the speed increment ΔV is the objective function.

For a monopropellant engine, the speed increment ΔV is:

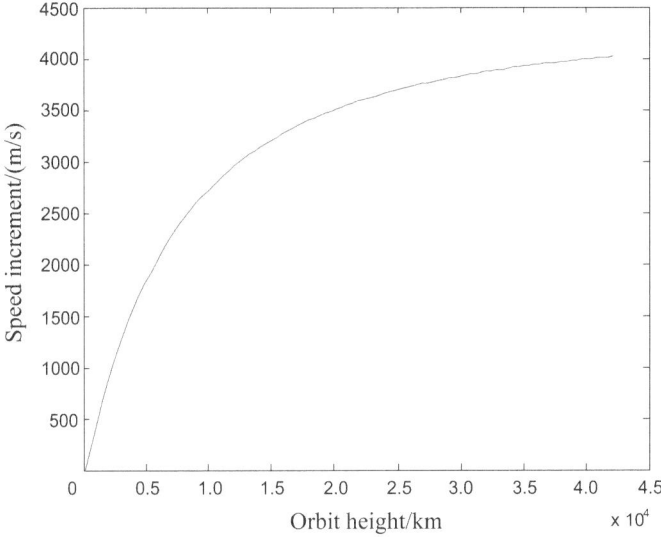

Fig. 8.9 Variation of speed increment ΔV with orbit height

$$\Delta V = I_{sp}g_0 \ln \frac{m_0}{m_f} \tag{8.4}$$

where m_0 is the initial mass of the spacecraft, m_f is the mass of the spacecraft after the propellant is used up and the orbit transfer is completed, I_{sp} is the specific impulse of the engine, and g_0 is the earth's gravitational acceleration.

When the Hohmann transfer orbit is used to start the orbit transfer from the initial circular orbit of 200 km, the variation of the speed increment needed for the satellite versus the orbit height is shown in Fig. 8.9. If the possible maximum ΔV is 1500 m/s, the transfer orbital height is 3700 km.

In the space transfer task, obtaining the maximum speed increment is the ultimate goal of task design. The specific impulse of the propulsion system I_{sp} and the final mass m_f of the spacecraft after the propellant is used up are the main factors affecting the speed increment. For an STP system, the specific impulse I_{sp} and mass m_f are comprehensively affected by parameters such as the mass of the tank, the collecting area of the concentrator and the mass of the thruster.

(2)   Payload mass

With the satellite's transfer orbit determined, the design objective is to obtain the maximum payload mass. In an STP system, the mass of the concentrator, the propellant mass and the mass of the tank are the main factors affecting the mass of the payload.

8.3.2.2   Design Variables

When implementing integrated design, only key parameters are selected as design variables in variable selection, which simplifies the model, reduces the computational volume, and avoids unnecessary interference by secondary factors. The main design variables are as follows:

(1) Thrust action time t_p

Thrust action time is the working time of the thruster in space, that is, the time from the ignition of the engine to the depletion of the propellant. The instantaneous mass of the spacecraft at time t is

$$m = m_0 - \frac{m_p}{t_p}t \tag{8.5}$$

(2) Collection area of the concentrator S_c

The collection area of the concentrator can affect the mass of the concentrator and the solar collection power. The propulsion power can be expressed as

$$P_T = \frac{1}{2}Fv_e = IS_c\eta \tag{8.6}$$

where I is the solar constant and η is the total efficiency of converting solar energy to propellant kinetic energy. The concentrator is a component that accounts for a large proportion of the mass of the STP system. For a space mission that requires N-level thrust, the collection area of the concentrator also needs to be on the order of several square meters. Increasing S_c can increase the solar collection power, which in turn can increase the specific impulse and thrust of the propulsion system, but at the same time, it increases the structural mass of the system. The area density of aluminum as the material of the concentrator is 24 kg/m^2, and the area density of the carbon fiber reinforced polymer (CFRP) is even lower at 11 kg/m^2.

(3) Propellant mass m_p

The propellant mass has a great impact on the speed increment; the larger the proportion of propellant mass in the total mass is, the larger the achievable speed increment. However, an increase in the propellant mass brings about an increase in the mass of the tank, and the two needs should be balanced in the design.

(4) Mass of the storage tank m_t

The mass of the tank is also the main structural mass of the STP system, and comprehensive consideration of many factors is needed, such as the propellant mass m_p and thrust action time t_p. For a spherical tank with the simplest structure, the mass is

$$m_t = 4\pi a^2 t_s \rho \tag{8.7}$$

where a is the nominal radius of the tank, ρ is the density of the structural material of the tank, and t_s is the wall thickness of the tank; $t_s = p_t a/(2S_w e_w)$. Using the radius of the tank as a variable, the mass of the tank and the propellant mass can be expressed as

$$m_t = \frac{2\pi a^3 p_t \rho}{S_w e_w} \tag{8.8}$$

$$m_p = \frac{4}{3}\pi a^3 \rho_p \tag{8.9}$$

where p_t is the maximum working pressure of the tank, S_w is the maximum allowable working stress of the structural material of the tank, and e_w is the welding seam efficiency.

(5) Specific impulse I_{sp}

The specific impulse is the parameter that has the largest impact on the spacecraft speed increment. Under different specific impulses, the variation in the needed propellant mass versus the speed increment is shown in Fig. 8.10. With a certain speed increment, a high specific impulse can reduce the requirement for propellant mass, thereby increasing the mass ratio of the payload. Both the specific impulse and thrust can be adjusted during orbital operation.

(6) Thrust F

Thrust is directly related to the specific impulse and propellant mass flow rate.

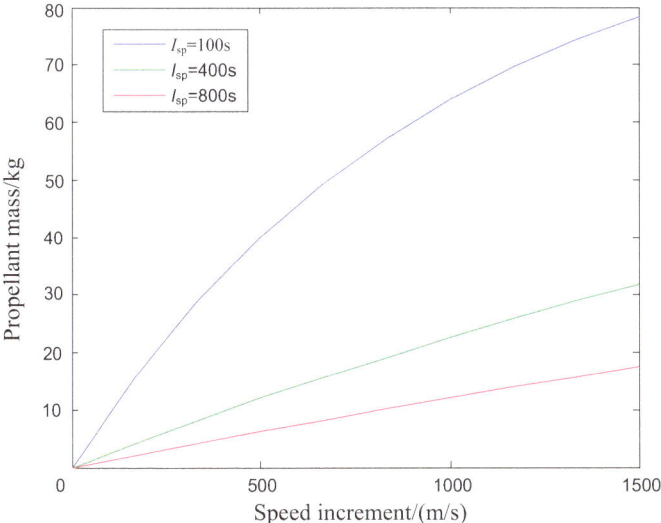

Fig. 8.10 Variation of propellant mass versus speed increment under different specific impulses

$$F = I_{sp}g_0\dot{m} = I_{sp}g_0\frac{m_p}{t_p} \tag{8.10}$$

Increasing thrust can reduce the propellant action time t_p, which reduces the time for tasks such as orbit transfer, but the increase in thrust requires the concentrator to provide greater solar power, which in turn needs to increase the mass of the concentrator, resulting in an increase in the structural mass of the system.

8.3.2.3 Constraints

(1) Value range of design variables

The specific impulse with the hydrogen as the propellant is 600–900 s, the specific impulse with the ammonia as the propellant is 200–600 s, the radius of the storage tank is 0.1–0.5 m, and the area of the concentrator is 1–5 m^2.

(2) Propulsion system mass constraint

When small satellites are used as the research object, the propellant mass should not exceed 50% of the total mass.

(3) Volume constraint

With small satellites as the research object, the volume of propellant tanks should not exceed 25% of the total volume of the satellite.

8.3.3 Analysis of Simulation Results

8.3.3.1 Hydrogen as a Propellant

Mission analysis is performed using liquid hydrogen as a propellant. Liquid hydrogen is a low-density and low-temperature propellant (boiling point 20 K), and very effective measures must be taken to control heat loss and prevent boiling vaporization.

First, the application in the 100 kg microsatellite is analyzed. With the speed increment as the objective function and the specific impulse, tank radius, and concentrator area as design variables, the evolutionary process by the quantum genetic algorithm is shown in Fig. 8.11. The speed increment of such small satellites using liquid hydrogen as a propellant is too small, only 316 m/s, and the orbit transfer task cannot be completed.

The use of polymer material for the concentrator can reduce the mass of the concentrator, and the speed increment can be increased to above 466 m/s, which is 15% higher than when using aluminum as the material, as shown in Fig. 8.12. Therefore, in a space task, the concentrator should be made of lightweight material. The main reason for the small speed increment of liquid hydrogen as a propellant is that the storage density is too low at only 71 kg/m^3, and the mass of liquid hydrogen

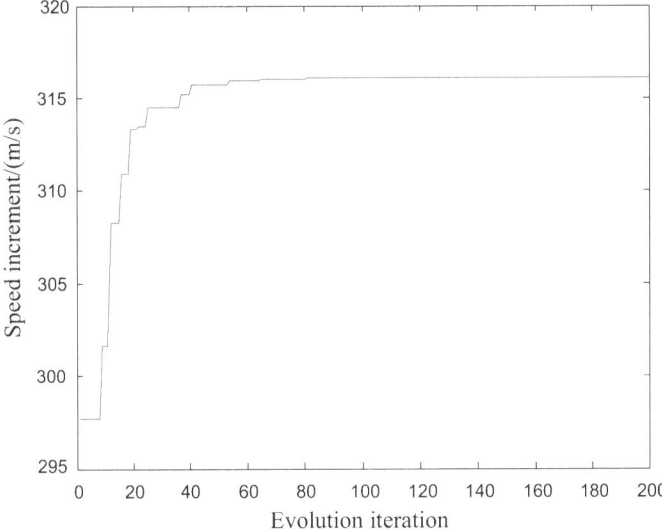

Fig. 8.11 Optimization process when using a liquid hydrogen propellant in a 100 kg small satellite

that can be carried in the space of a small satellite is too small at only 5.2 kg, which is far less than the limit of propellant mass able to be carried by small satellites, but the occupied volume reaches the allowable volume upper limit.

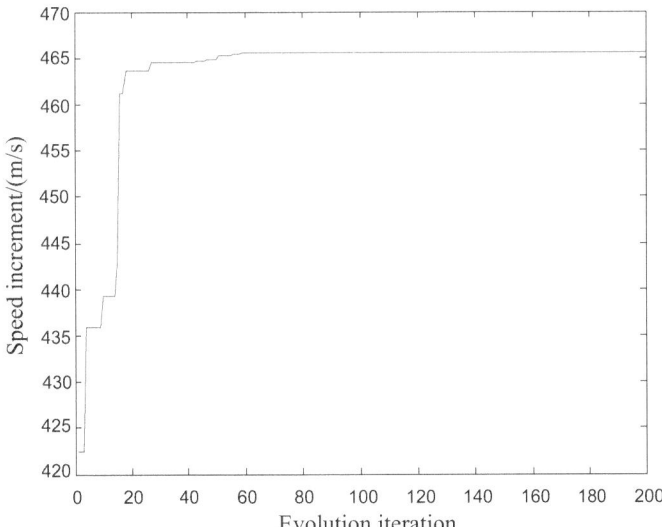

Fig. 8.12 Task optimization process using a polymer concentrator

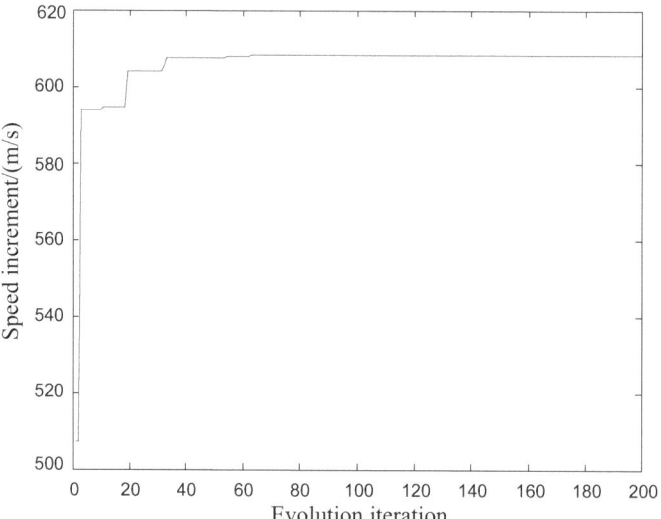

Fig. 8.13 The application of a liquid hydrogen propellant to a 400 kg small satellite

For a 400 kg small satellite space transfer task, the speed increment when using a liquid hydrogen propellant is increased to 608 m/s, as shown in Fig. 8.13. As the total mass and volume of the spacecraft increase, the liquid hydrogen propellant can gradually meet the mission requirements, as shown in Fig. 8.13.

The optimal value is obtained when the specific impulse is 800 s and the tank radius is 0.4 m. Although the increase in tank volume causes an increase in the mass of the tank and the propellant mass at the same time, the increase in propellant mass has a greater impact on the speed increment, which cancels the decreasing trend caused by the increase in the mass of the tank. When the radius of the tank is set to the maximum, the maximum speed increment can be obtained.

The ideal design for the use of a liquid hydrogen propellant is to increase the obtained volume proportion of the tank in the satellite by reducing the volume of other components. If the volume proportion of the liquid hydrogen tank is increased to 50%, the speed increment can reach 1143 m/s, which can satisfy various orbit transfer and orbit adjustment tasks for a satellite, as shown in Fig. 8.14. However, the multilayer heat insulation of the tank and the subcooling ventilation system needed to maintain the low temperature of liquid hydrogen are not considered in the analysis, but the mass and volume proportions of this part are relatively large. Therefore, the direct application of liquid hydrogen requires improvements in cryogenic fluid storage technology.

Liquid hydrogen propellants can be used in large satellites or other spacecrafts. The satellite in the Integrated Solar Upper Stage (ISUS) program in the United States is taken as the research objects, with a total mass of 3630 kg and a designed thrust of 6 N. An optimization is performed for this task. The quantum evolution process shown in Fig. 8.15 shows that a speed increment of 6730 m/s can be obtained.

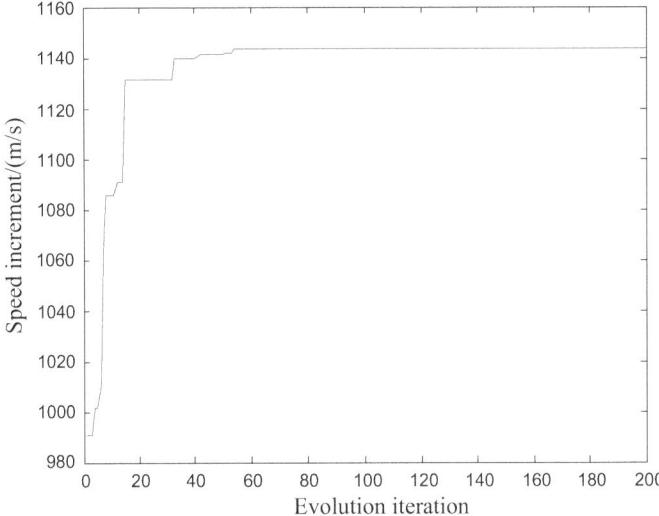

Fig. 8.14 The case when the liquid hydrogen storage tank accounts for 50% of the volume

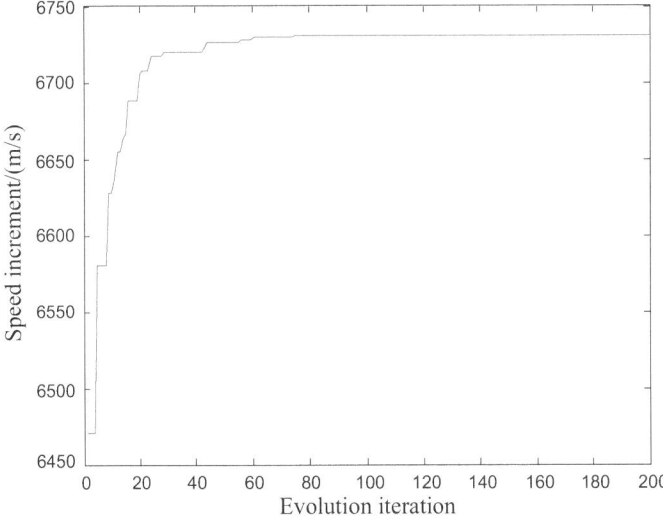

Fig. 8.15 Design optimization for the application of liquid hydrogen in large satellites

8.3.3.2 Ammonia as a Propellant

Under the same design conditions, the speed of a 100 kg microsatellite using liquid ammonia as a propellant reaches 2676 m/s when the CFRP concentrator is used, as shown in Fig. 8.16. Liquid ammonia has a higher storage density (600 kg/m^3),

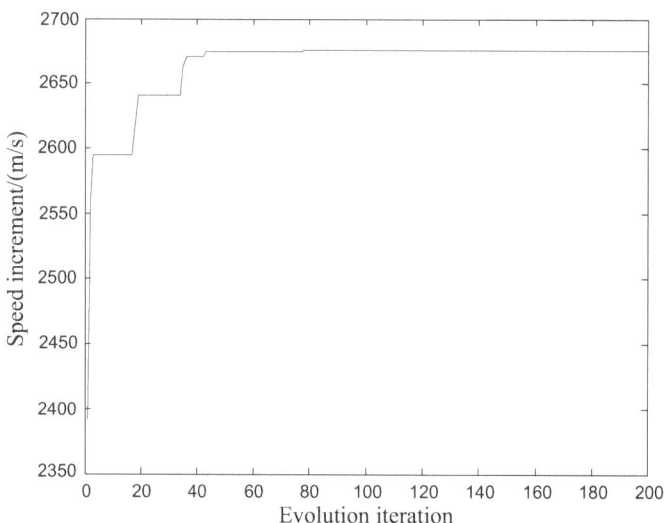

Fig. 8.16 Liquid ammonia propellant under the 25% volume limit for a 100 kg small satellite

and under the same volume limitation, liquid ammonia can better meet the needs of satellites for space missions.

For the situation in which the propellant volume on the satellite is limited to 17%, the use of ammonia as a propellant can also obtain a speed increment of 1970 m/s, as shown in Fig. 8.17. The performance when using ammonia for STP in a small satellite is very high.

When the 400 kg small satellite uses ammonia as a propellant, the acquired speed increment can reach 4100 m/s, as shown in Fig. 8.18, and the orbit transfer task from LEO to GEO can be directly completed.

With the maximum payload as the optimization goal, the payload optimization process of a 100 kg small satellite is shown in Fig. 8.19 when the speed increment is fixed at 1500 m/s and the specific impulse, propellant mass, and condenser mass are used as design variables.

In the payload optimization, the total efficiency of the small satellite propulsion system has a great impact on the optimization results. Table 8.2 shows the optimization results under different total efficiencies. High efficiency can reduce the system's demand for solar power, increase the specific impulse of the propulsion system, and reduce the concentrator mass and the propellant mass.

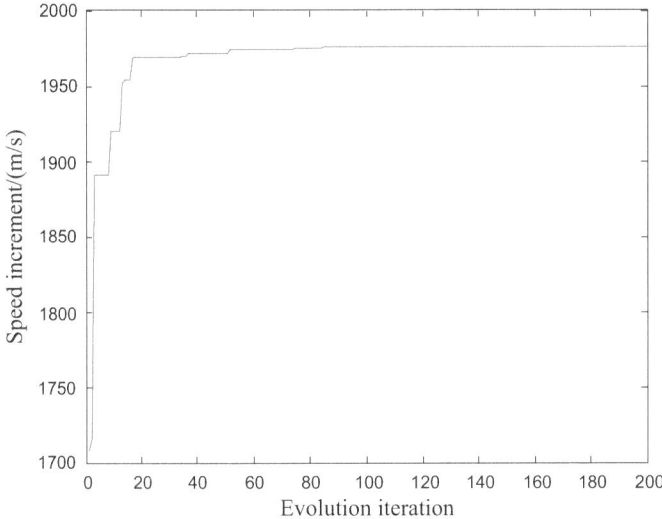

Fig. 8.17 Liquid ammonia propellant under the 17% volume limit for a 100 kg small satellite

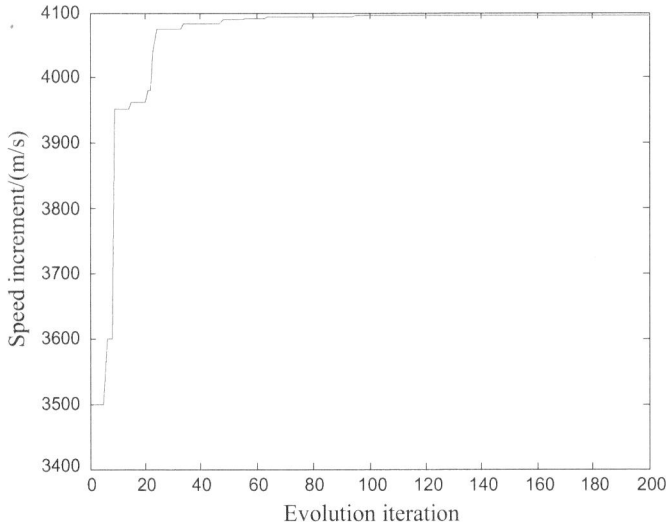

Fig. 8.18 Liquid ammonia propellant in a 400 kg small satellite

Figure 8.20 shows the variation curves of payload mass versus specific impulse under different efficiencies. There is an optimal payload, and it is not the case that the larger the specific impulse, the larger the payload mass, which is due to the fact that as the specific impulse increases, the concentrator area needs to keep increasing

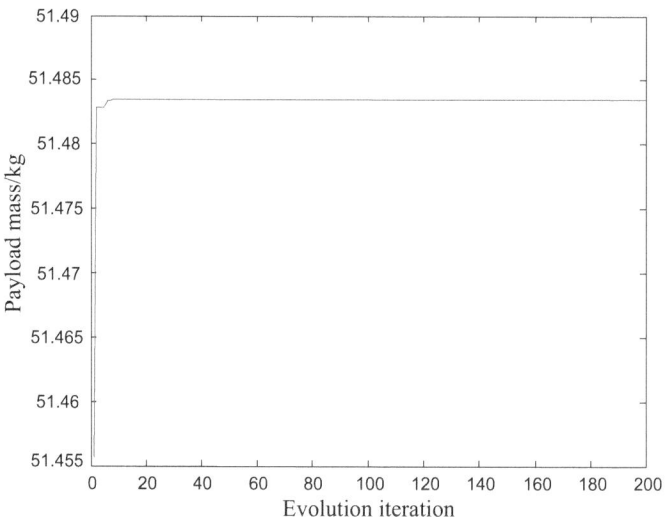

Fig. 8.19 Payload optimization process of a 100 kg small satellite

Table 8.2 Optimization results for small satellites under different total efficiencies of propulsion systems

Total efficiency	Specific impulse/s	Propellant mass/kg	Concentrator mass/kg	Payload mass/kg
0.9	508	26.0	22.4	51.5
0.7	438	29.5	24.8	45.6
0.5	355	35.1	28.1	36.7

to maintain the solar power required by the system, thus bringing about an increase in the structural mass of the system and reducing the mass proportion of the payload.

8.4　Comprehensive Analysis of Thruster Performance

The variations in the working gas mass flow rate, the temperature of the heated working fluid, the speed of the working fluid at the nozzle exhaust, the thrust of the thruster, and the specific impulse are shown in Fig. 8.21.

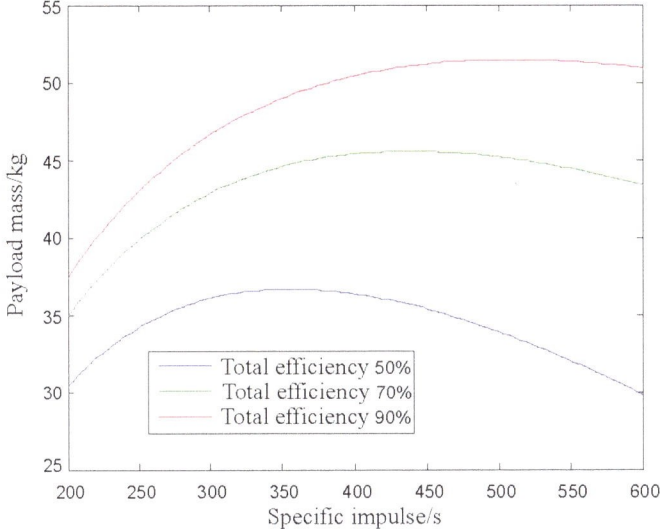

Fig. 8.20 Variation curve of payload mass versus specific impulse under different efficiencies

Figure 8.21 shows that when the mass flow rate is lower, the working gas can obtain a higher temperature before entering the nozzle, and thus the calculated speed and specific impulse at the nozzle exhaust are larger. However, if the mass flow rate is too small, it cannot meet the thrust design requirement for the entire system. When the mass flow rate changes within the range of $0.5 \times 10^{-4} - 2 \times 10^{-4}$ kg/s, the changes in gas entry temperature, nozzle exhaust speed, thrust, and specific impulse are approximately linear, which provides great convenience for the adjustment of related parameters. In addition, previous studies have found that increasing the nozzle expansion ratio can also improve the specific impulse of the thruster, but the increasing effect is no longer significant when the expansion ratio reaches 50:1.

Fig. 8.21 Relationship curves for working gas performance changes

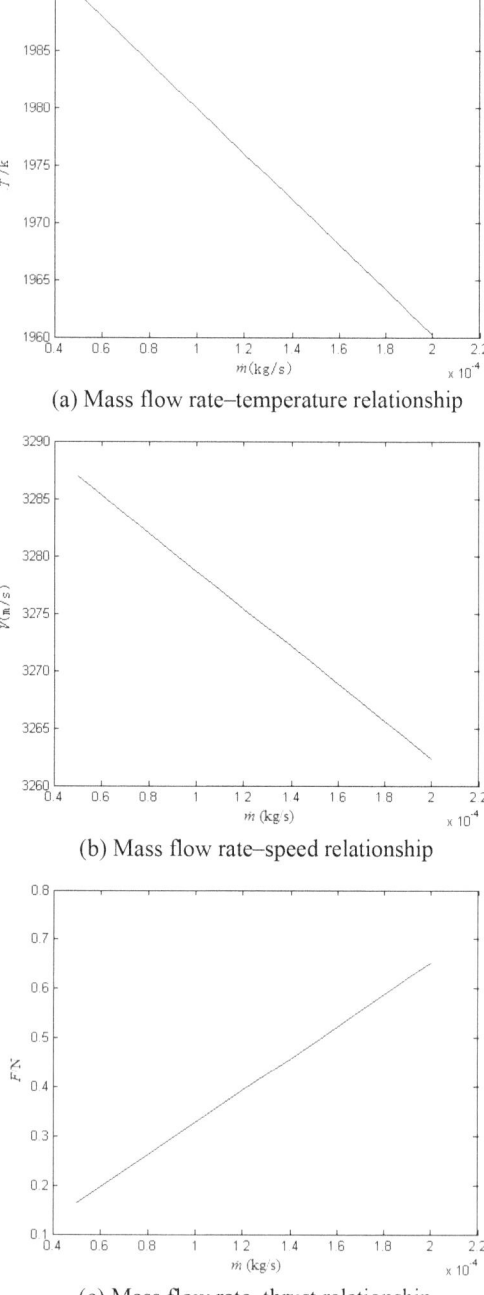

(a) Mass flow rate–temperature relationship

(b) Mass flow rate–speed relationship

(c) Mass flow rate–thrust relationship

Fig. 8.21 (continued)

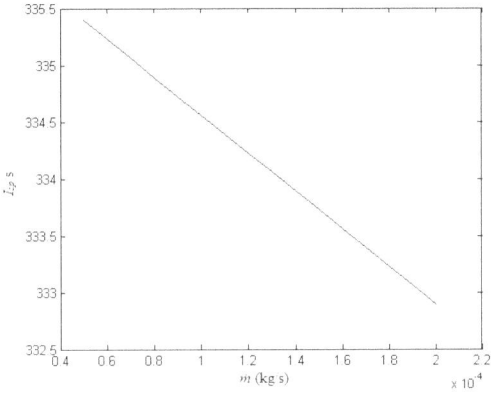

(d) Mass flow rate–specific impulse relationship

Figure 8.21 shows that when the mass flow rate is 1.6×10^{-4} kg/s, the thrust can reach 0.5 N; when the mass flow rate is 0.6×10^{-4} kg/s, the thrust can reach 0.2 N. Table 8.3 shows the relationship between the total impulse and the propellant mass under the above two conditions.

The following two points can be seen in Table 8.3: (1) when the total impulse is 1000–100,000 N s and the propellant mass is appropriate, STP has a great advantage; (2) when the total thrust requirement is the same, increasing the propellant by 6.7% can increase the thrust by 150%.

Table 8.3 Relationship between total impulse and propellant mass under two working conditions

Design thrust/N		0.5	0.2
Mass flow rate/(kg/s)		1.6×10^{-4}	0.6×10^{-4}
Total impulse 1000 N s	Working time/s	2000	5000
	Propellant mass/kg	0.32	0.3
Total impulse 10,000 N s	Working time/s	20,000	50,000
	Propellant mass/kg	3.2	3
Total impulse 100,000 N s	Working time/s	200,000	500,000
	Propellant mass/kg	32	30
Total impulse 1,000,000 N s	Working time/s	2,000,000	5,000,000
	Propellant mass/kg	320	300

Chapter 9
Performance Validation Experiment for the Solar Thermal Thruster

9.1 Introduction

The secondary concentrator and thrust chamber of the solar thermal thruster are integratedly designed using the regenerative cooling method, and the heat exchanger core of the solar thermal thruster is designed using the laminate structure. Based on theoretical analysis and numerical simulation results and with the help of a solar thermal propulsion (STP) experimental system, an experimental study of the optimized design of solar thermal thrusters is carried out.

9.2 STP Experiment System

The STP experiment system used in the experimental study is composed of a propellant supply system, a thruster tester, a vacuum chamber and a vacuum pump, a xenon lamp light source system, and a thruster. The system is shown in Fig. 9.1.

9.2.1 Propellant Supply System

The propellant supply system is mainly composed of nitrogen cylinders, pressure reducing valves, gas storage cylinders, pressure gauges, globe valves, and pipelines. During use, nitrogen is depressurized from the steel cylinder through the pressure reducing valve and then enters the four carbon fiber-wound gas cylinders for inflation, in order to simulate the working state of the space storage tank. The propellant supply system is shown in Fig. 9.2.

© National University of Defense Technology Press 2025   151
M. Huang et al., *Solar Thermal Thruster*, https://doi.org/10.1007/978-981-97-7490-6_9

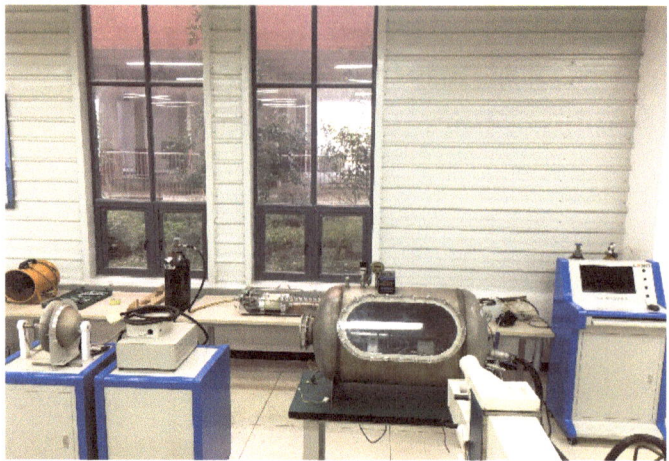

Fig. 9.1 STP experiment system

Fig. 9.2 Propellant supply system

9.2.2 Thruster Tester

The thruster tester is mainly composed of a data acquisition card, various sensors, data wires and corresponding acquisition software, as shown in Fig. 9.3. The data acquisition sensors of the experimental system include a propellant volume flow sensor, thruster temperature sensor, thruster pressure sensor, and nozzle exhaust thrust sensor. The specifications and models of each piece of equipment on the tester are shown in Table 9.1.

The thrust sensor is a piezoresistor that measures the generated thrust by measuring the reaction force of the propellant on the force-receiving surface of the sensor after the propellant is ejected from the nozzle, with an acquisition accuracy of 0.01 N, as shown in Fig. 9.4. The temperature is measured by a K-type thermocouple and displayed by a multicircuit temperature recorder, with an acquisition accuracy of 0.1 °C, as shown in Fig. 9.5.

Fig. 9.3 Tester for thruster with a small thrust

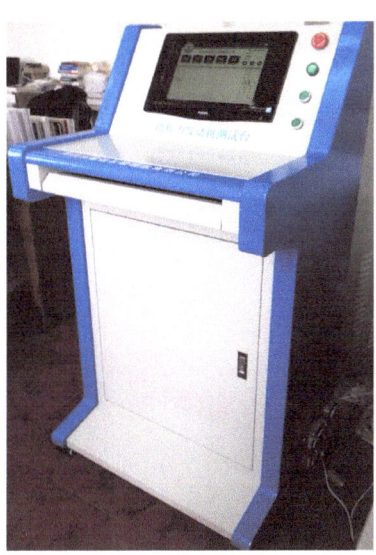

Table 9.1 Specifications and models of thrust tester

Name	Model/specification	Measuring range	Remarks
Industrial control cabinet	$60 \times 60 \times 160$ cm	–	–
Industrial computer	Lenovo	–	1 piece
Data acquisition instrument	DAQ-0800	–	1 piece
N_2 flow sensor	SFC4200	0–20 SLM	1 measuring point
Pressure sensor	YSZK-313	0–4 MPa	1 measuring point
Temperature sensor	K-type	– 200 to 1370 °C	11 measuring points
Thrust sensor	BK-3A	0–10 N	1 measuring point
Temperature recorder	90 series	–200–1370 °C (K-type)	1 piece
	Combined instrument SH-X	–200–1370 °C (K-type)	1 piece

9.2.3 Xenon Lamp as the Light Source

Since the xenon lamp spectrum is basically consistent with the solar spectrum, solar radiation is simulated by using a xenon lamp as the light source. The radiation spectral energy distribution of the xenon lamp is close to that of sunlight, with a color temperature of approximately 6000 K. The spectral distribution of the continuous spectrum part is almost independent of the input power of the xenon lamp, and the spectral energy distribution almost does not change during the lifetime. The light

Fig. 9.4 Thrust sensor

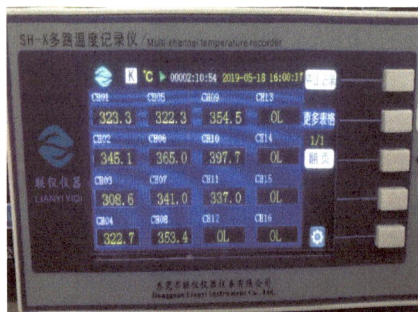
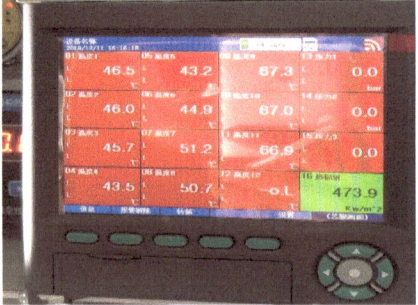

(a) Combined instrument SH-X   (b) 90 series

Fig. 9.5 Temperature recorder

source system is composed of a xenon lamp, a primary concentrator, a power control box, and wires. The output power of xenon lamp, the light source, ranges from 700 to 7000 W, and the efficiency of converting electrical power to light energy is approximately 60%. During the experiment, the output power of the light source is adjusted by the knob of the power control box (the power is gradually increased to prevent the rupture of the secondary concentrator caused by the large thermal shock). The light is converged by the primary concentrator and then incident on the surface of the secondary concentrator of the thruster. The light source is shown in Fig. 9.6.

9.2.4 Flow Controller

In the STP experiment system, the flow of the propellant nitrogen is changed through the flow controller. The working principle of the flow controller is to change the cross-sectional area of the pipeline through the input voltage, thus achieving flow control, and the flow control accuracy is 0.01 nL/min (20 °C, 101.325 kPa). The controller

(a) Light source system before the experiment

(b) Light source system during the experiment

Fig. 9.6   Xenon lamp as the light source

is shown in Fig. 9.7. When the input voltage is 0, the flow controller is off, and the nitrogen flow rate is 0. When the input voltage is greater than or equal to 5 V, the flow controller is fully opened, and the nitrogen flow rate reaches the maximum value of 20.00 nL/min.

With the pressure of the gas cylinder being constant at 0.4 MPa, the flow rate of nitrogen gas under different input voltages is obtained through experiments, and the results are shown in Table 9.2.

The voltage of the flow controller is plotted against the corresponding flow rate, and the variation pattern is shown in Fig. 9.8. Figure 9.8 shows that the relationship between the flow rate and the voltage is basically linear. Within the working pressure range of the flow controller, the flow rate can be basically determined by the applied voltage.

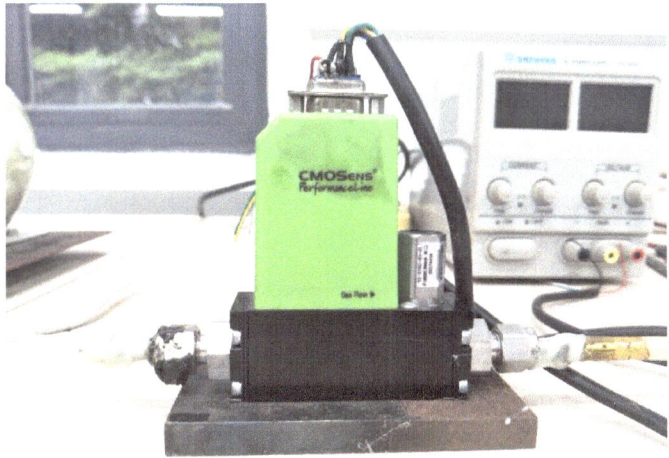

Fig. 9.7   Flow controller

Table 9.2 Experimental data of the nitrogen flow rate under different voltages at 0.4 MPa

Voltage/V	Flow rate/(nL/min)	Voltage/V	Flow rate/(nL/min)	Voltage/V	Flow rate/(nL/min)
0	0	1.8	7.41	3.6	14.53
0.2	1.16	2.0	8.19	3.8	15.20
0.4	1.75	2.2	8.90	4.0	16.04
0.6	1.73	2.4	9.89	4.2	16.70
0.8	3.51	2.6	10.50	4.4	17.77
1.0	4.17	2.8	11.15	4.6	18.37
1.2	5.01	3.0	12.31	4.8	19.01
1.4	5.75	3.2	13.02	5.0	20.00
1.6	6.71	3.4	13.64		

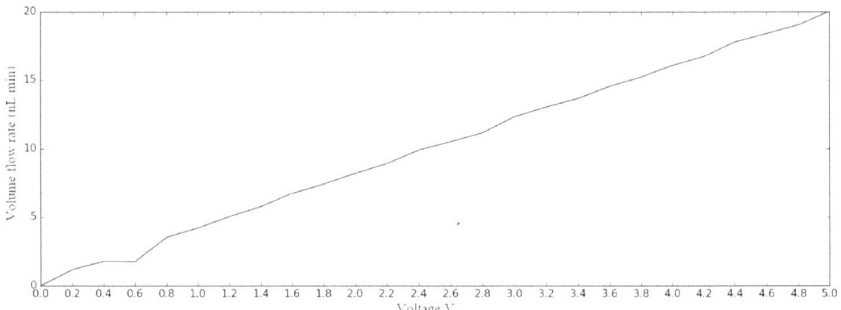

Fig. 9.8 Relationship between flow rate and voltage of flow controller at 0.4 MPa

At the same time, experiments are conducted on the flow controller under nitrogen pressures of 0.3 and 0.4 MPa, as shown in Fig. 9.9. Figure 9.9 shows that the nitrogen flow rate is unstable when the pressure is low (0.3 MPa), while the nitrogen flow rate is relatively stable when the pressure is 0.4 MPa.

9.3 Cold Gas Propulsion Experiment of Thruster

According to the principle and experimental needs of STP, an experimental system is built, the propellant gas pipelines are connected, the data acquisition sensor is connected and debugged, and the cold gas propulsion of a thruster is carried out.

The thruster is installed in a stainless steel sleeve, which is installed in a vacuum chamber together with the thrust sensor and pressure sensor, as shown in Fig. 9.10 a and b. The temperature of the thruster is measured by 11 K-type thermocouples at the front (3), middle (3), rear (3) and throat (2) of the thruster. The parameter

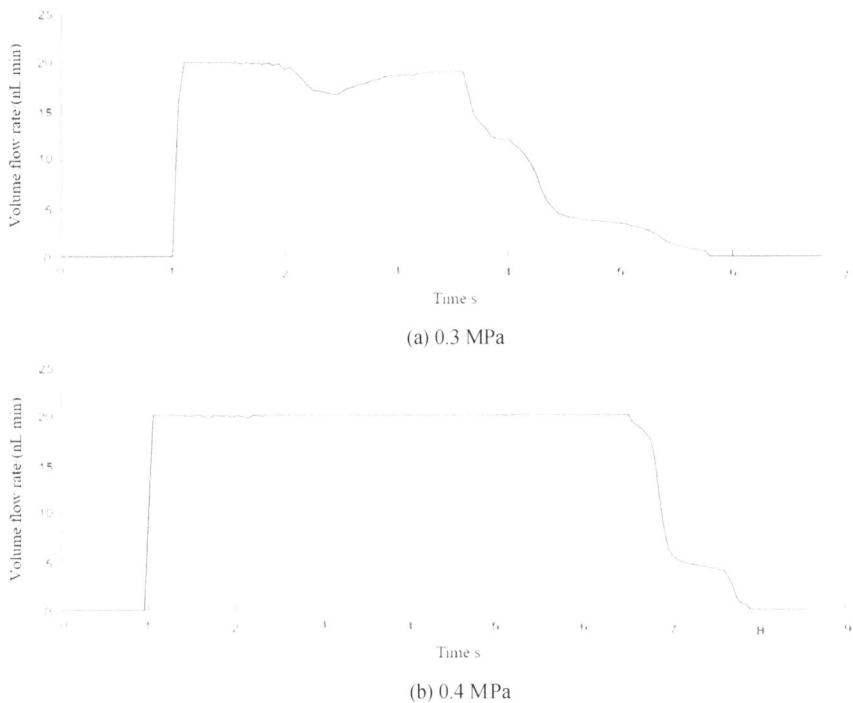

(a) 0.3 MPa

(b) 0.4 MPa

Fig. 9.9 Comparison of flow rate stability under different pressures

recorder is installed outside the vacuum chamber. The distribution of the thruster wall temperature is shown in Fig. 9.10c and d. The flow sensor is installed outside the vacuum chamber. The vacuum chamber is evacuated by an external vacuum pump to reduce the air pressure. Currently, the minimum measured pressure in the chamber is 12.8 Pa.

The pressure of the gas cylinder is 0.4 MPa. The cold gas test is conducted on the thruster under three working conditions: working condition 1, working condition 2, and working condition 3. Based on the nitrogen flow rates corresponding to different voltages in Table 9.2, the different working conditions are shown in Table 9.3.

The experimental data obtained under cold gas tests (two groups of data for each working condition) are plotted in Fig. 9.11, and the thrust is shown in Table 9.4.

Under the same gas cylinder pressure of 0.4 MPa and unheated propellant, the corresponding thrust is measured by adjusting the propellant flow rate. In Fig. 9.11, the solid blue line represents the propellant flow rate, and the dotted red line represents the thrust of the thruster. The experimental data show that after the propellant valve is opened, the flow rate rises rapidly to the maximum value: for working condition 1 (voltage 5 V), the propellant flow rate basically remains unchanged after reaching the maximum value; for working conditions 2 (voltage 4 V) and 3 (voltage 3 V), the propellant flow rate decreases rapidly after reaching the maximum value and then

(a) Thruster fixing sleeve

(b) Installation of thruster and thrust sensor in the chamber

(c) Temperature collection points at the front and middle of the thruster

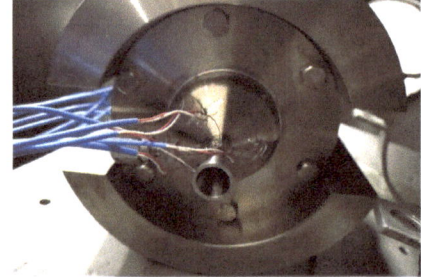

(d) Temperature collection points at the rear and throat of the thruster

Fig. 9.10   Thruster installation and thrust and temperature data acquisition

Table 9.3 Propellant flow rates under different working conditions

Working condition	1	2	3
Voltage/V	5.0	4.0	3.0
Propellant volume flow rate/(nL/min)	20.0	16.5	12.5

stabilizes around a certain value, indicating that the flow controller has a better control effect on the propellant flow rate. Under the three working conditions, the obtained thrust quickly reaches the maximum after a short delay and then gradually decreases. The reason is that after the propellant is ejected from the nozzle, the vacuum in the vacuum chamber decreases rapidly, which increases the resistance as propellant is ejected from the nozzle; simultaneously, the thrust sensor has limited accuracy, and the error in the measured value increases over time. Therefore, the maximum value of the thrust curve is taken as the measured nozzle thrust.

It should be noted that for nitrogen, the theoretical specific impulse of gas at room temperature is 75 s, and the actual measurement value is between 60 and 72 s.

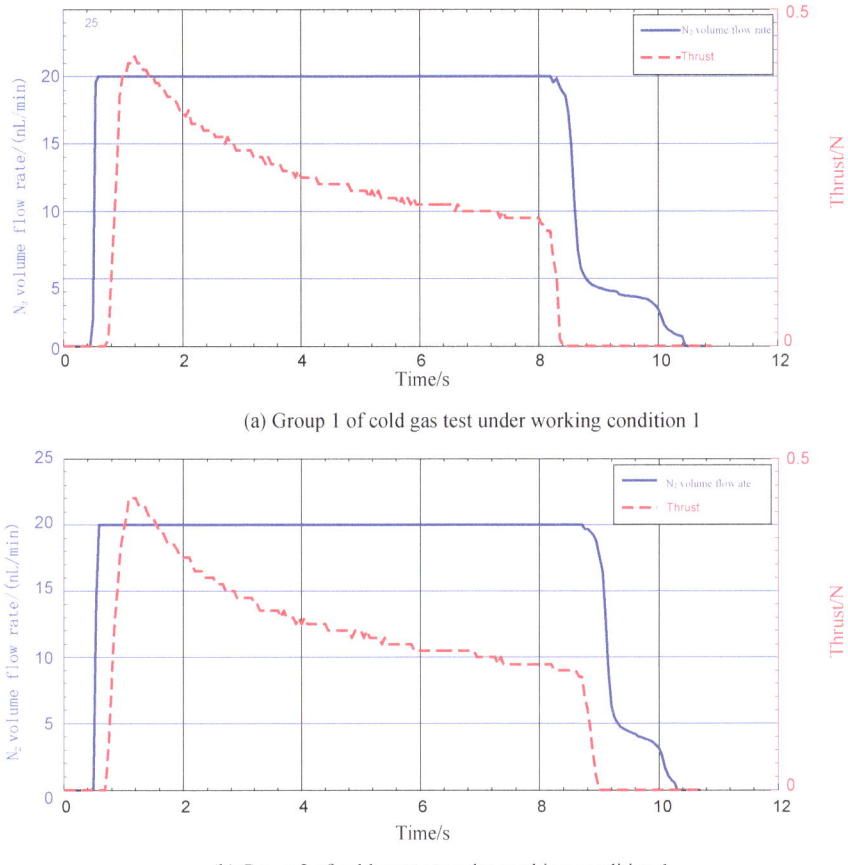

(a) Group 1 of cold gas test under working condition 1

(b) Group 2 of cold gas test under working condition 1

Fig. 9.11   Variation curves of propellant flow rate and thrust with the unheated propellant

9.4   Thruster Propellant Heating Experiment

In the experiments, a strong xenon lamp light source is used to simulate solar radiation. Since the spectrum of the xenon lamp is basically consistent with the solar spectrum, the color temperature is approximately 6000 K. The electric power of the light source ranges from 700 to 7000 W. The efficiency of converting electrical power to light energy is approximately 60%. During the experiment, the power of the light source is gradually increased to prevent the secondary concentrator of the thruster from being damaged or even ruptured by the large thermal shock. Figure 9.12 shows the xenon lamp heating test and the condition inside the chamber during heating, and Fig. 9.13 shows the melting of the heat insulation sleeve and temperature measuring wire after the heating test.

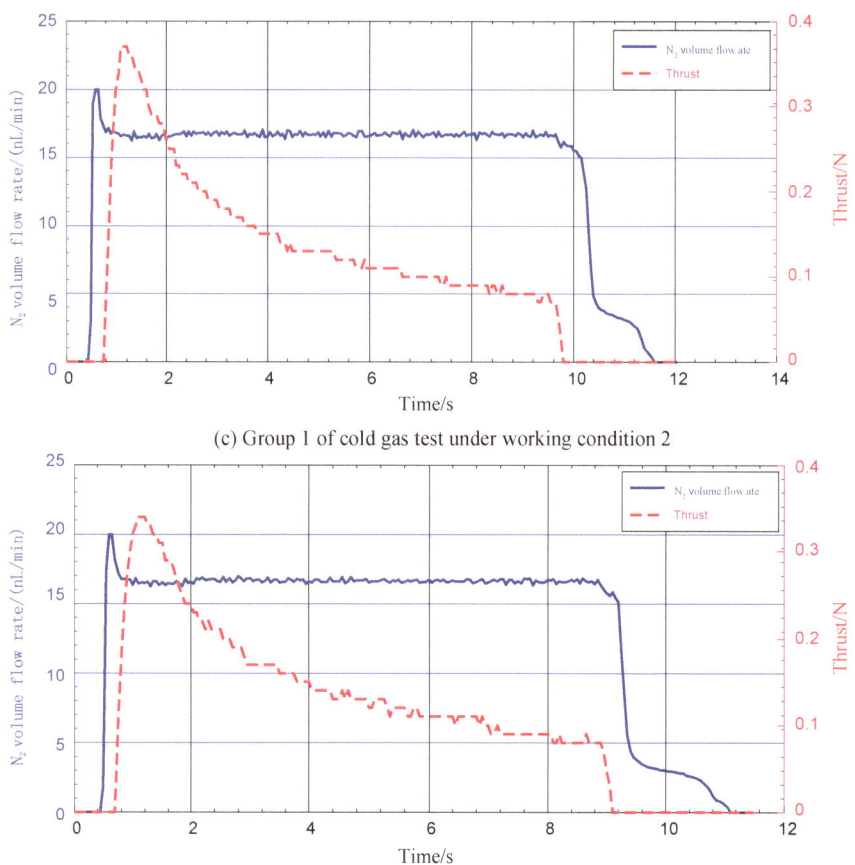

(c) Group 1 of cold gas test under working condition 2

(d) Group 2 of cold gas test under working condition 2

Fig. 9.11 (continued)

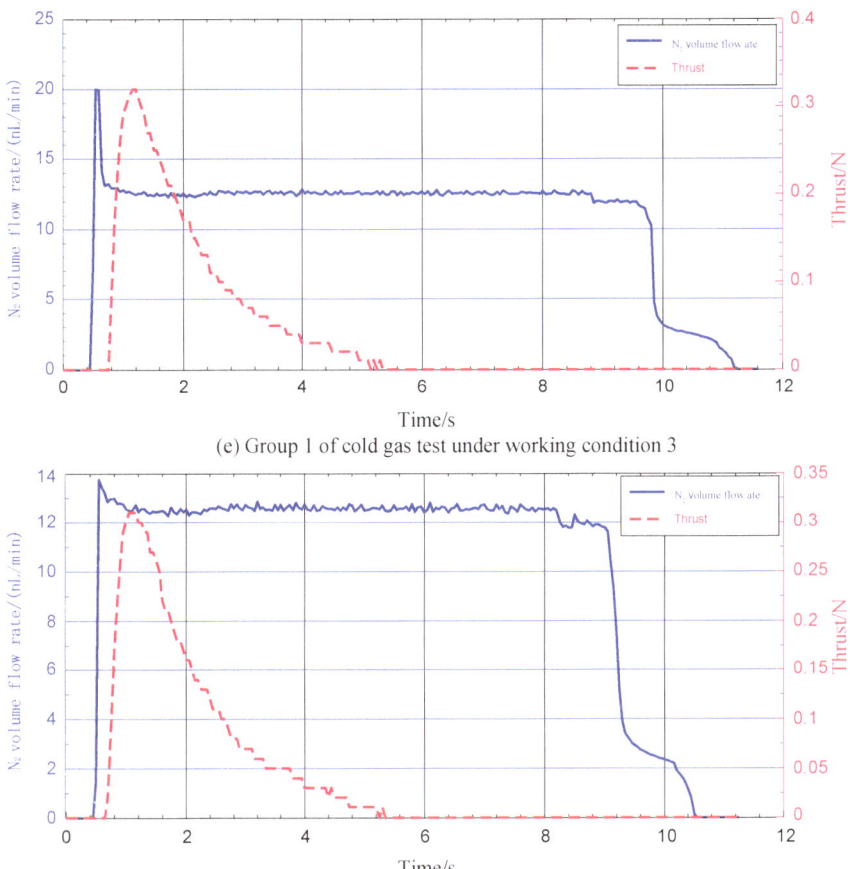

(e) Group 1 of cold gas test under working condition 3

(f) Group 2 of cold gas test under working condition 3

Fig. 9.11 (continued)

Table 9.4 Thrusts Corresponding to Different Conditions

Working condition	1		2		3	
Propellant volume flow rate/ (nL/min)	20.0		16.5		12.5	
Propellant mass flow rate/($\times 10^{-4}$ kg/s)	7.0737		5.8358		4.4210	
Thrust/N	Group 1	Group 2	Group 1	Group 2	Group 1	Group 2
	0.43	0.44	0.37	0.37	0.32	0.31
Specific impulse/s	60.79	62.20	63.40	63.40	72.38	70.12

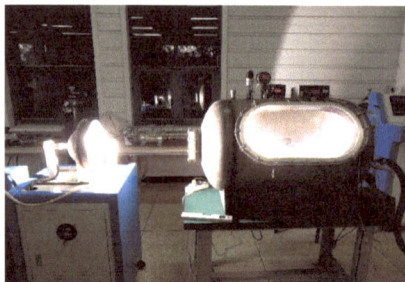
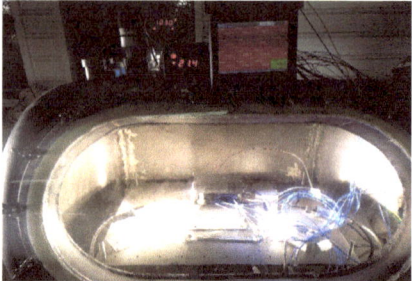

Fig. 9.12 Xenon lamp heating experiment and inside condition

Fig. 9.13 The heat insulation sleeve and temperature measuring wire melted after the heating experiment

In the early heating experiments, the outer surface of the thruster is exposed to high temperature for a long period of time, which causes the melting and sticking of the enameled wire outside the temperature measuring wire. To prevent the enameled wire outside the temperature measuring wire in the high-temperature conditions to melt, thus causing a short-circuit of inner metal wire to affect the temperature measurement, in the later experiments, the wires near the thruster are wrapped with asbestos sheets to avoid the exposure of the temperature-measuring wires to strong light, as shown in Fig. 9.14.

9.4.1 The First Heating Experiment

The gas cylinder pressure is maintained at 0.4 MPa, and the propellant flow rate is maintained at that of working condition 2 (approximately 16.5 nL/min). Propellant heating experiments are conducted on a thruster under seven xenon lamp powers of 1000, 2000, 3000, 4000, 5000, 6000 and 7000 W, and the measured temperatures corresponded to the power of the xenon lamp are 60, 100, 150, 200, 250, 300, and 350 °C.

Fig. 9.14   Asbestos-coated temperature measuring wires

The experimental data obtained under different xenon lamp powers as heating conditions are plotted in Fig. 9.15, the nitrogen density is shown in the Table 9.5, and thrusts are shown in the Table 9.6.

The internal temperature of the thrust chamber can be calculated by the following formula (temperature calculation results are for reference only):

$$\frac{T_1}{T_0} = \left(\frac{I_1}{I_0}\right)^2$$

where T_0 represents the temperature of the outer wall of the thruster; T_1 represents the internal temperature of the thrust chamber; I_1 represents the specific impulse of the thruster when the temperature of the outer wall of the thruster is T_0; and I_0 represents the specific impulse of the thruster when the temperature of the outer wall of the thruster is room temperature. The corresponding thrust chamber temperature is calculated based on the different outer wall temperatures and specific impulses of the thruster. The calculation results are shown in Table 9.7.

As shown in Fig. 9.13, after the heating test, the wall surface of the thruster and the heat insulation sleeve all suffer different degrees of burning and discoloration; the blue enameled wire of the temperature measuring wire is melted and adhered, and the secondary concentrator is intact.

The experimental curves show that the flow rate is basically stable at 16.5 nL/ min during the whole experimental process. As the xenon lamp power continues to increase, the thrust generated by the thruster gradually increases, and the specific impulse gradually rises. When the power reaches 6000 W (300 °C), the thrust reaches the maximum value of 0.59 N; when the light source power continues to increase, the thrust of the thruster basically remains unchanged, and the specific impulse continues to rise. When the xenon lamp power is 7000 W, the thrust can reach 0.58 N and the specific impulse can reach 131.40 s when nitrogen is used as the propellant. The specific impulse of the thruster can be increased to a certain extent by increasing the incident power, and the variation trend and increase are shown in Fig. 9.16 and Table 9.8.

Figure 9.16 shows that the thrust and specific impulse of the thruster with a light source power $\leq$ 6000 W increase linearly with increasing temperature, and the increments in thrust and specific impulse are approximately 0.00032 N/W and 0.0114 s/W, respectively. The thrust measured when the light source power is 7000 W is the same as that when the power is 6000 W. The reason may be that the gap between the force-receiving surface of the thrust sensor and the nozzle exhaust is slightly increased, resulting in the measured thrust being slightly lower than the actual thrust.

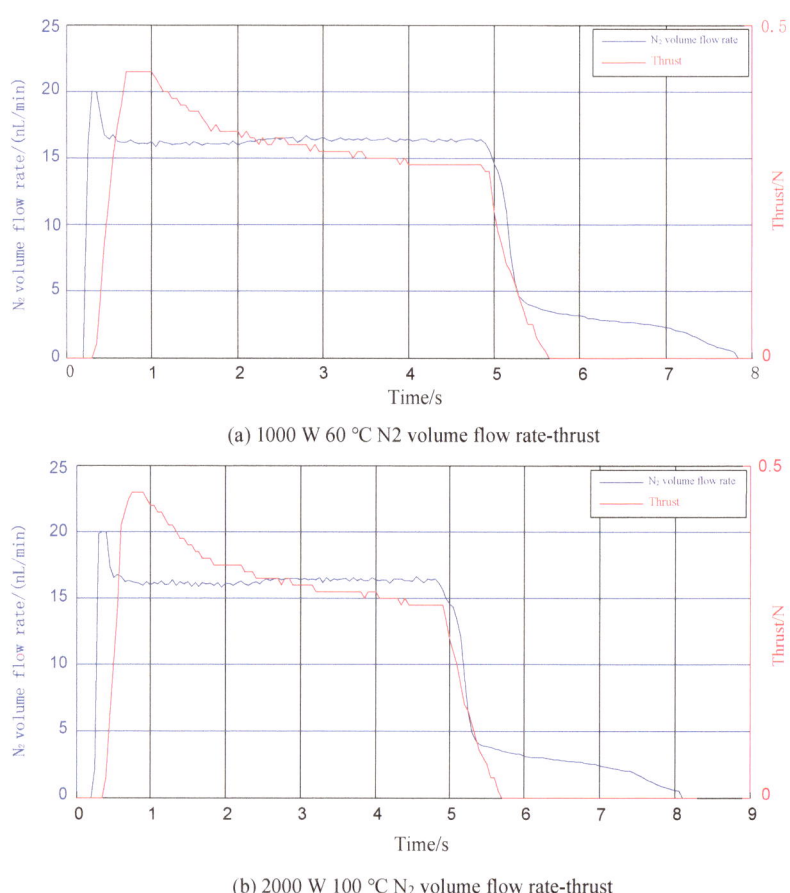

(a) 1000 W 60 °C N2 volume flow rate-thrust

(b) 2000 W 100 °C N$_2$ volume flow rate-thrust

Fig. 9.15 Variations of thrust under different xenon lamp powers in thruster propellant heating experiment

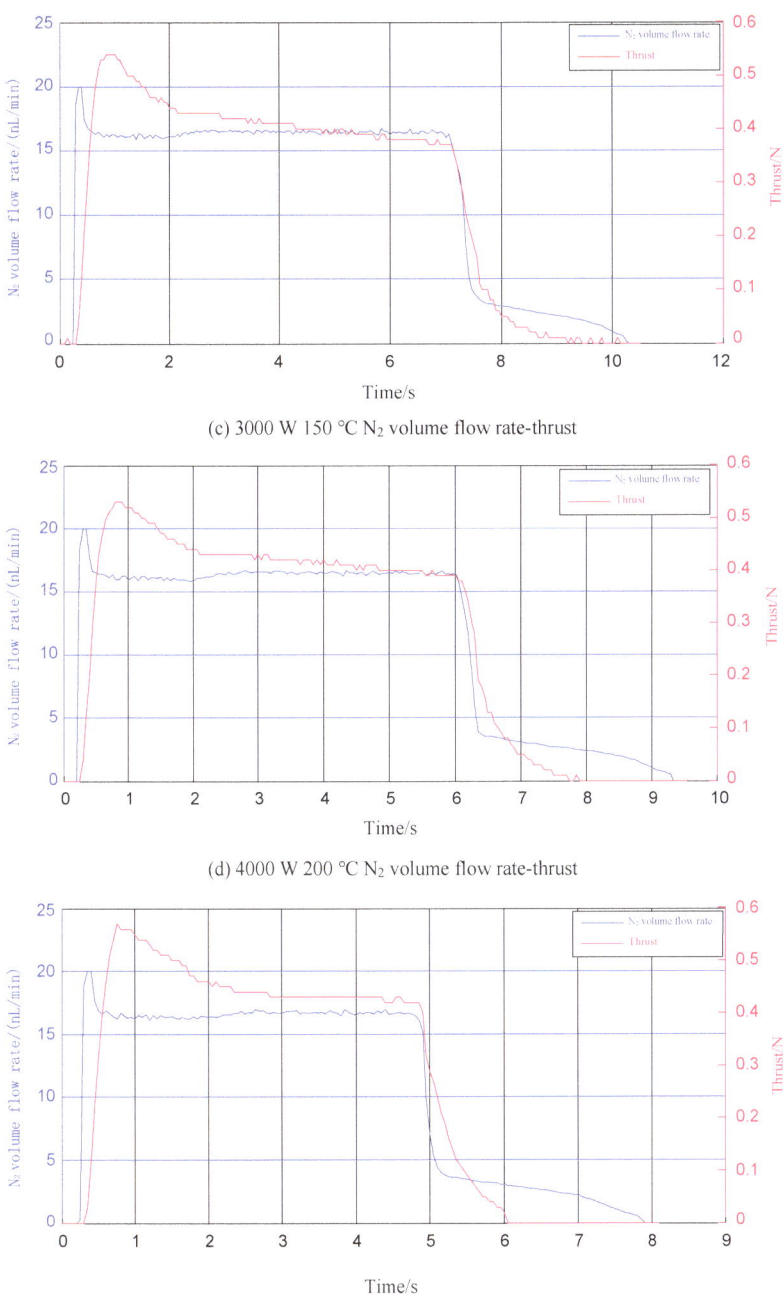

(c) 3000 W 150 °C N_2 volume flow rate-thrust

(d) 4000 W 200 °C N_2 volume flow rate-thrust

(e) 5000 W 250 °C N_2 volume flow rate-thrust

Fig. 9.15 (continued)

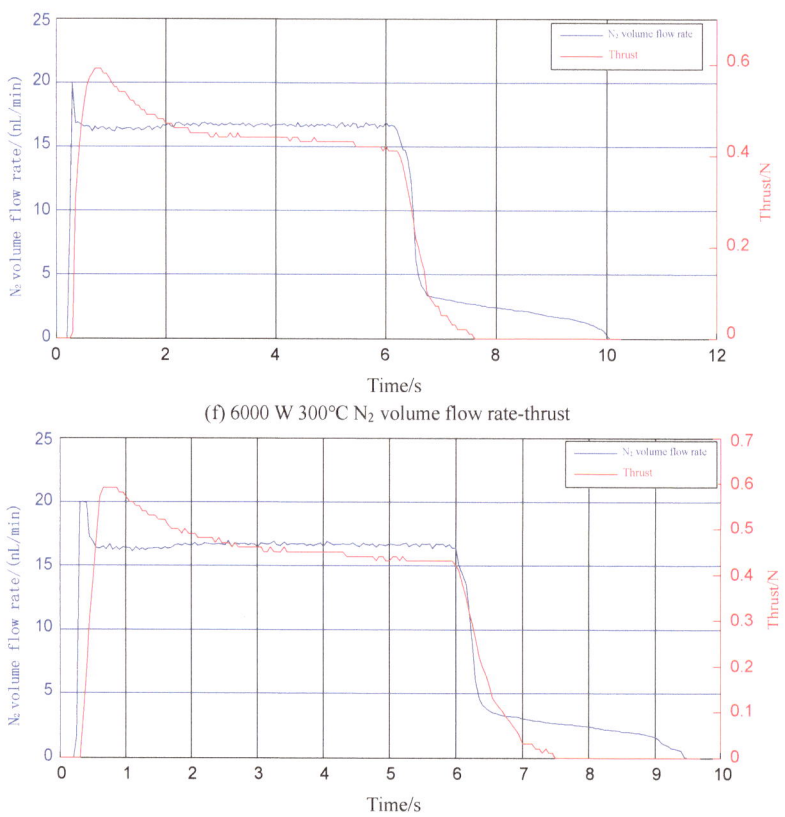

(f) 6000 W 300°C N₂ volume flow rate-thrust

(g) 7000 W 350 °C N₂ volume rate flow-thrust

Fig. 9.15   (continued)

Table 9.5   Nitrogen density under different powers

Power/W	1000	2000	3000	4000	5000	6000	7000
Outer wall temperature/°C	60	100	150	200	250	300	350
Pressure/MPa	0.246	0.255	0.268	0.276	0.285	0.293	0.297
Density/(kg/m³)	2.4868	2.3015	2.1330	1.9645	1.8347	1.7217	1.6051

9.4.2   The Second Heating Experiment

According to the first heating test results, the propellant flow rate is reduced. Under the same light source system, with the pressure of the gas cylinder at 0.4 MPa and the flow rate of the propellant at approximately 16 nL/min, the thrust tests are conducted on the thruster under nine thruster outer wall temperatures: 20 (room temperature), 60, 100, 150, 200, 250, 300, 350 and 400 °C.

Table 9.6 Thrusts corresponding to different xenon lamp powers

Xenon lamp power/ W	1000	2000	3000	4000	5000	6000	7000
Propellant volume flow rate/(nL/min)	16.5	16.5	16.5	16.5	16.5	16.5	16.5
Propellant mass flow rate/($\times 10^{-4}$ kg/s)	6.8387	6.3291	5.8657	5.4025	5.0454	4.7346	4.4141
Thrust/N	0.43	0.46	0.54	0.55	0.57	0.59	0.58
Specific impulse/s	62.80	72.68	92.06	101.80	112.97	124.61	131.40

Table 9.7 Thrust chamber temperatures corresponding to different outer wall temperatures

Power/W	1000	2000	3000	4000	5000	6000	7000
Outer wall temperature/°C	60	100	150	200	250	300	350
Propellant temperature/°C	–	–	232.3	426.1	699.7	1064.1	1419.8

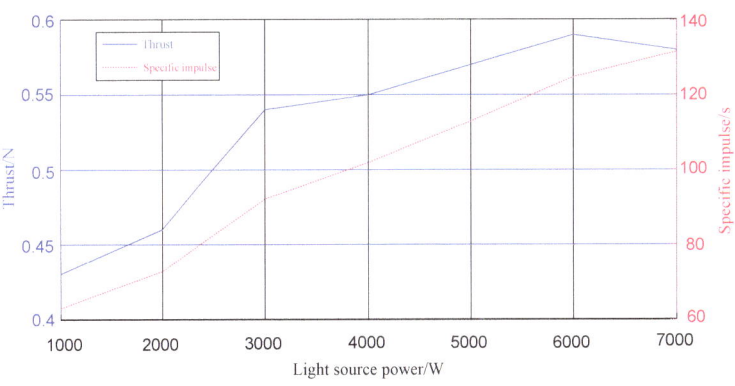

Fig. 9.16 Variations of thrust/specific impulse for thruster under different xenon lamp powers

Table 9.8 Improvement in specific impulses under different light source powers

Power/W	1000	2000	3000	4000	5000	6000	7000
Specific impulse/s	62.80	72.68	92.06	101.80	112.97	124.61	131.40
Specific impulse increment/s	0	9.88	29.26	39.00	50.17	62.81	68.60
Percent/%	0	15.73	46.59	62.10	79.89	98.42	109.24

The procedure of the heating experiment is shown in Fig. 9.17. The experimental data obtained under different thruster outer wall temperatures are plotted in Fig. 9.18, the nitrogen density at different temperatures is shown in Table 9.9, and the thrust is shown in Table 9.10.

The experimental temperature versus the thrust and specific impulse are plotted in Fig. 9.19. Figure 9.19 shows that at ≤ 400 °C, the thrust and specific impulse of the thruster increase linearly with increasing temperature, with increments in thrust and specific impulses of approximately 0.00052 N/°C and 0.21 s/°C, respectively. When nitrogen is used as a propellant, the thrust reaches 0.64 N, and the specific impulse reaches 148.5 s when the ambient temperature is 400 °C.

Compared with the those at room temperature, the thrust and specific impulse of the thruster show a greater increase with increasing temperature, and the increase of the specific impulse is approximately (10–16%)/50 °C. Table 9.11 shows the variation trend and increase of specific impulses.

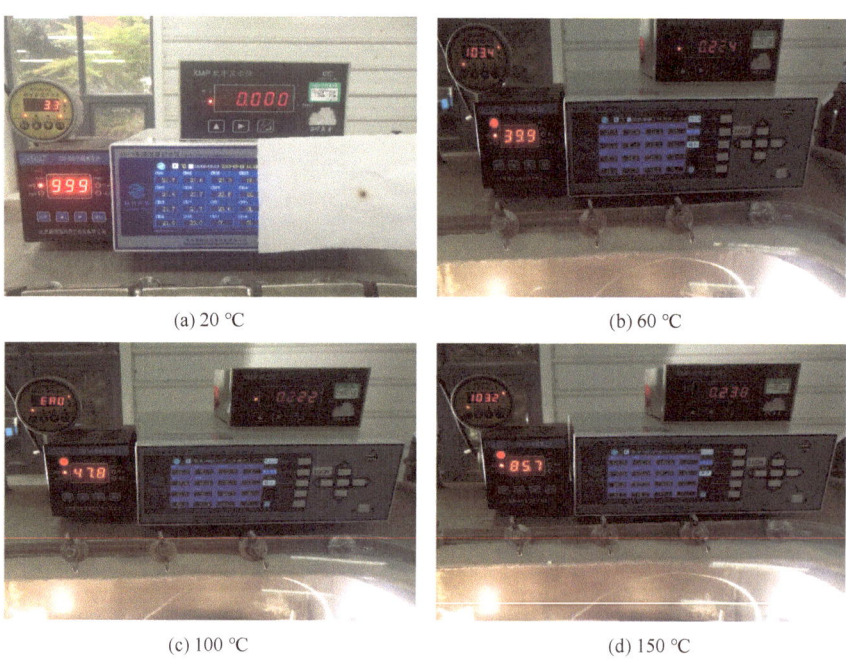

(a) 20 °C   (b) 60 °C

(c) 100 °C   (d) 150 °C

Fig. 9.17 Heating experiment procedure

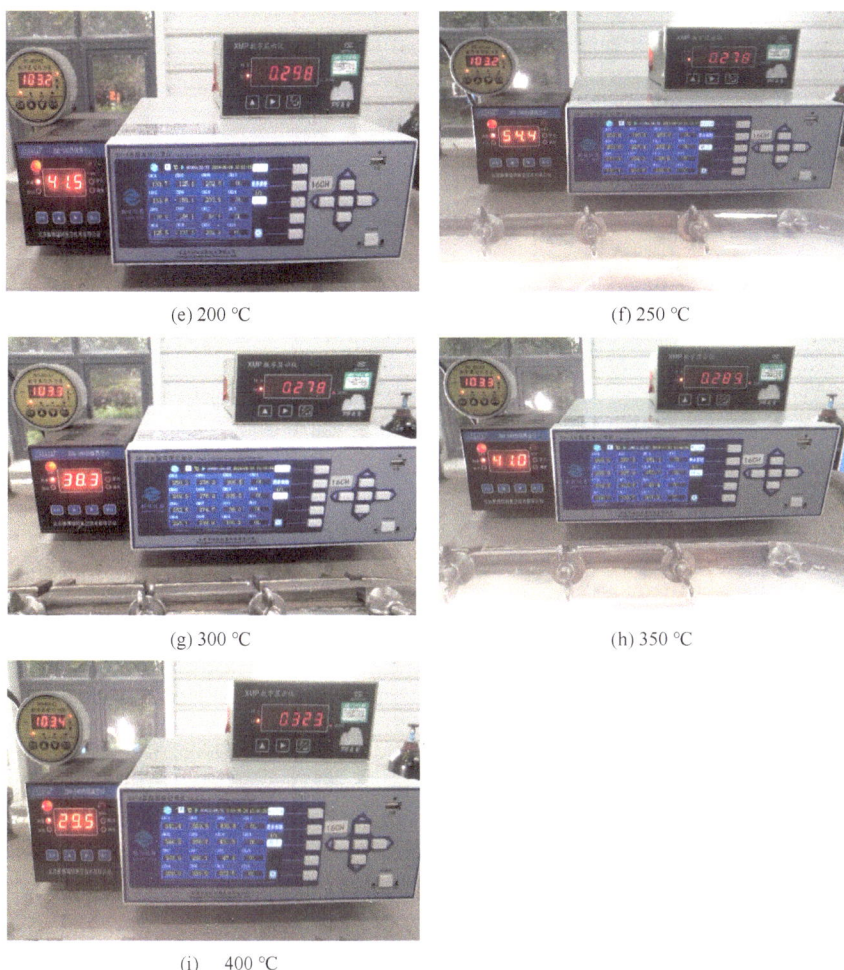

(e) 200 °C   (f) 250 °C

(g) 300 °C   (h) 350 °C

(i)   400 °C

Fig. 9.17 (continued)

After the heating test, the thruster is burned and discolored, as shown in Fig. 9.20. Figure 9.20 shows that after the heating test, the thruster wall suffers different degrees of burning and discoloration along the axial direction, the discoloration of the thruster head is the most severe, and the blue enameled wire of the temperature measuring wire is melted and adhered, but the secondary concentrator is intact. The temperature of the thruster is distributed in the order of front > middle > nozzle throat > nozzle expansion.

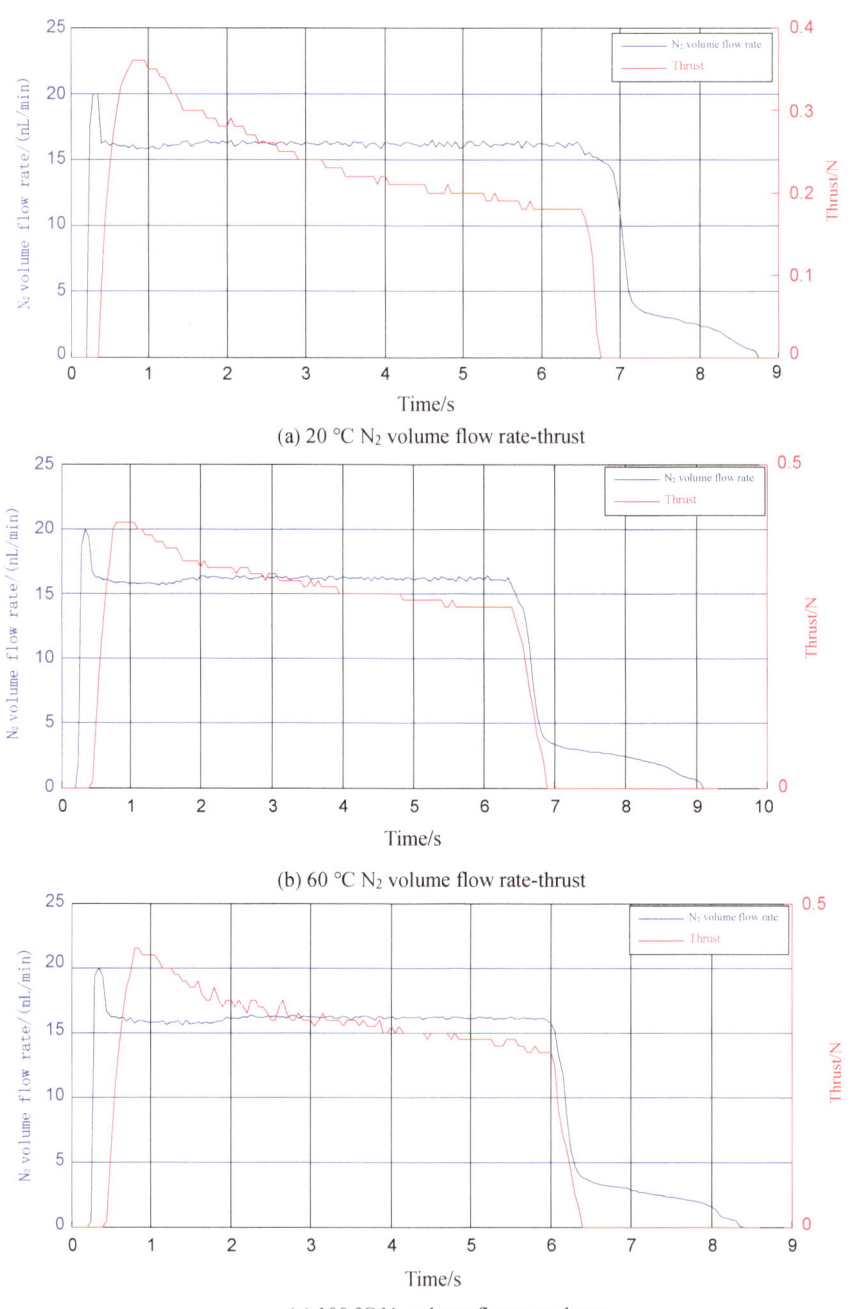

(a) 20 °C N₂ volume flow rate-thrust

(b) 60 °C N₂ volume flow rate-thrust

(c) 100 °C N₂ volume flow rate-thrust

Fig. 9.18 Variations of N₂ volume flow rate versus thruster thrust under different thruster outer wall temperatures

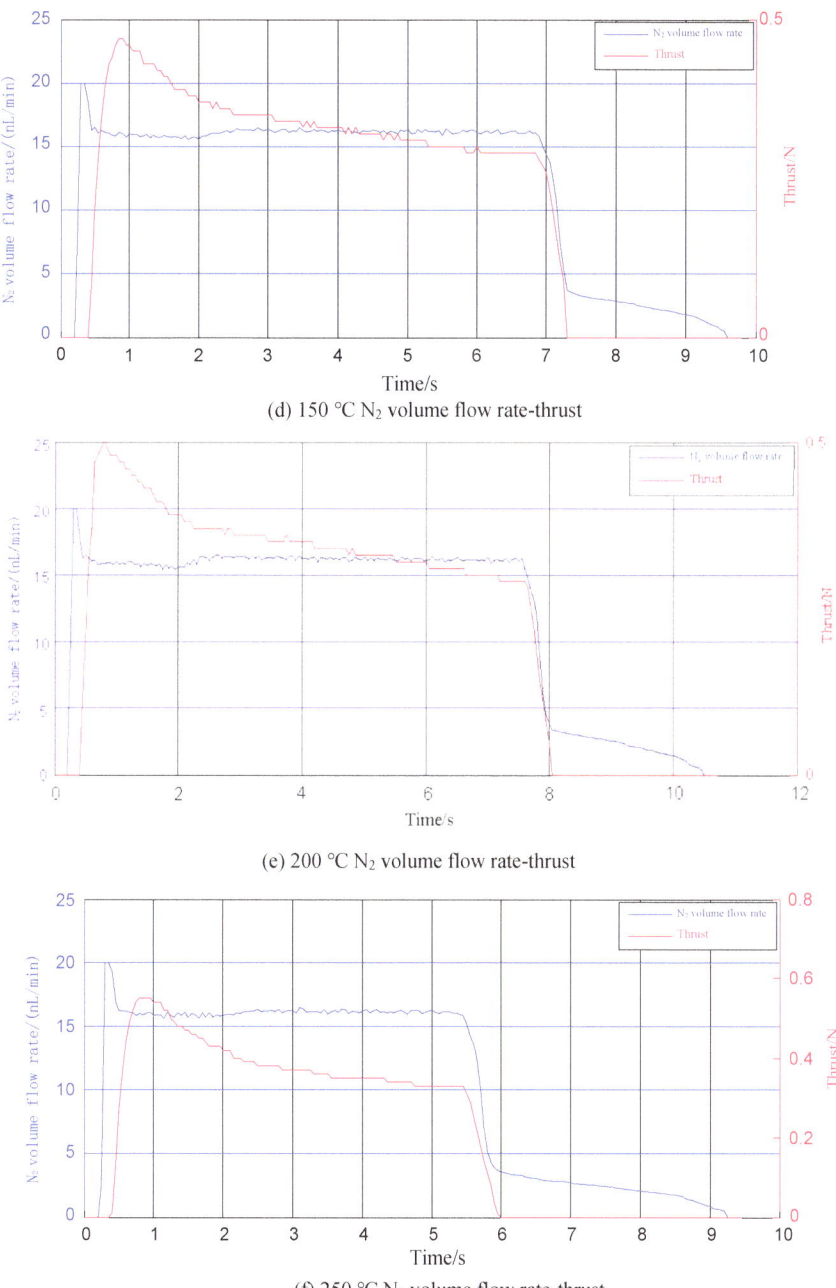

(d) 150 °C N_2 volume flow rate-thrust

(e) 200 °C N_2 volume flow rate-thrust

(f) 250 °C N_2 volume flow rate-thrust

Fig. 9.18 (continued)

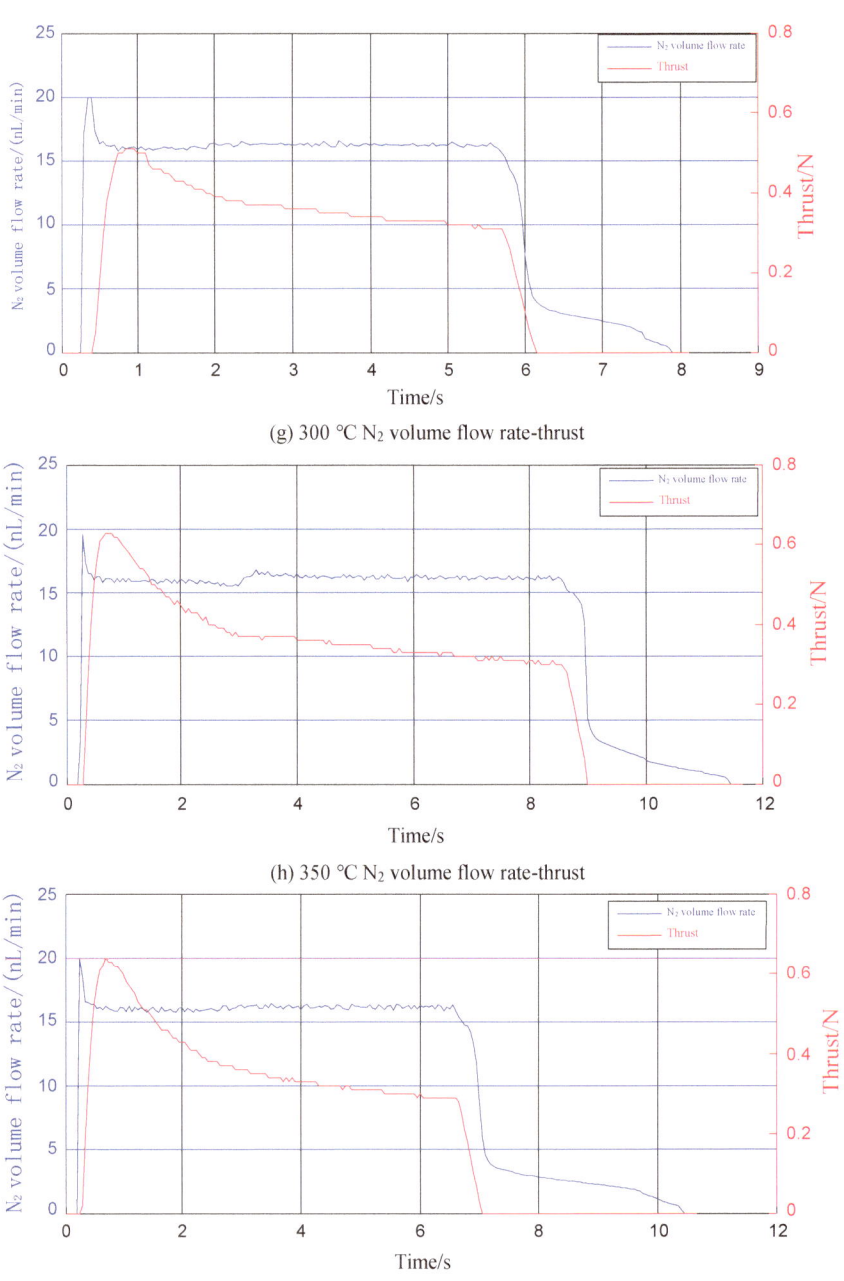

(g) 300 °C N₂ volume flow rate-thrust

(h) 350 °C N₂ volume flow rate-thrust

(i) 400 °C N₂ volume flow rate-thrust

Fig. 9.18 (continued)

Table 9.9   Corresponding nitrogen densities at different temperatures

Temperature/°C	20	60	100	150	200	250	300	350	400
Pressure/MPa	0.210	0.222	0.237	0.242	0.251	0.271	0.283	0.303	0.323
Density /(kg/m^3)	2.412	2.244	2.139	1.926	1.787	1.745	1.663	1.637	1.616

Table 9.10   Thrusts corresponding to different thruster outer wall temperatures

Temperature/°C	20	60	100	150	200	250	300	350	400
Propellant volume flow rate/ (nL/min)	16.0	16.0	16.0	16.0	16.0	16.0	16.0	16.0	16.0
Propellant mass flow rate/($\times$ 10^{-4} kg/s)	6.433	5.984	5.704	5.136	4.765	4.652	4.434	4.367	4.309
Thrust/N	0.33	0.41	0.43	0.47	0.50	0.55	0.57	0.62	0.64
Specific impulse/s	51.30	68.52	75.39	91.51	104.9	118.2	128.6	142.0	148.5

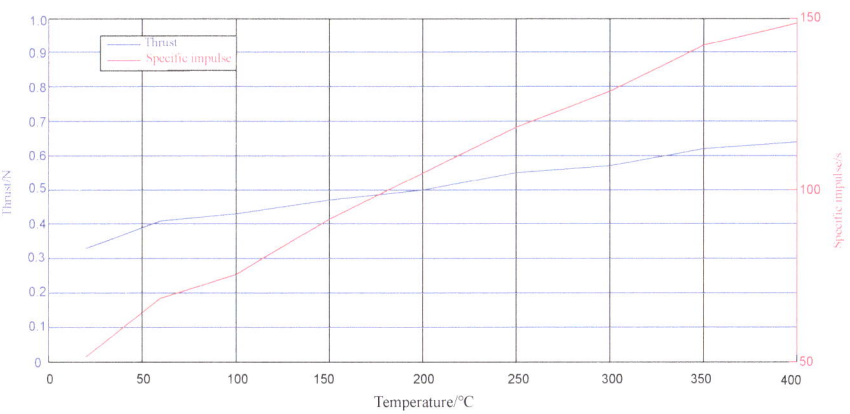

Fig. 9.19   Thrust and specific impulse of thruster under different outer wall temperatures

Table 9.11   Increases in specific impulse under different thruster outer wall temperatures

Temperature/°C	20	60	100	150	200	250	300	350	400
Specific impulse/s	51.30	68.52	75.39	91.51	104.9	118.2	128.6	142.0	148.5
Specific impulse increment/s	0	17.22	24.09	40.21	53.6	66.9	77.3	90.7	97.2
Percent/%	0	33.57	46.96	78.38	104.48	130.41	150.68	176.80	189.47

Fig. 9.20 Burning and discoloration of the thruster after the heating test

Chapter 10
Thrust of Solar Thermal Thruster Under Variable Working Conditions

10.1 Introduction

Experiments are conducted on the STP system built in Chap. 9 to test the thrust of the solar thermal thruster under different working conditions. In this chapter, the heating power and propellant flow rate are changed to change the thrust.

10.2 Cold Gas Propulsion Experiment of Thruster Under Variable Working Conditions

The experiments show that when the vacuum chamber is not evacuated or the vacuum of the vacuum chamber is low, the thrust cannot be experimentally measured due to the accuracy of the thrust sensor and the influence of the residual atmosphere. After the vacuum chamber is evacuated, the minimum pressure inside the chamber is approximately 12.8 Pa, and a cold gas test under different working conditions can be performed.

In the tests of the supply system pressure being 0.4, 0.5, 0.6, 0.7 and 0.8 MPa, the flow controller (when the input voltage is 5.0 V) is fully opened, and results of the cold gas experiment are shown in Fig. 10.1. Figure 10.1 show that at various pressures, the flow rates under these conditions are basically the same at 20 nL/min, suggesting that the flow controller can better control the propellant flow rate when the pressure is greater than or equal to 0.4 MPa.

As shown in Fig. 10.1, the solid line shows the measured propellant flow rate, and the dotted line shows the measured thrust. After the propellant valve is opened, the flow rate rises rapidly to reach the maximum value and then basically remains unchanged. The measured thrust quickly reaches the maximum value after a short delay and then gradually decreases. The reason is that after the propellant is ejected from the nozzle, the vacuum in the vacuum chamber decreases rapidly, which

© National University of Defense Technology Press 2025
M. Huang et al., *Solar Thermal Thruster*, https://doi.org/10.1007/978-981-97-7490-6_10

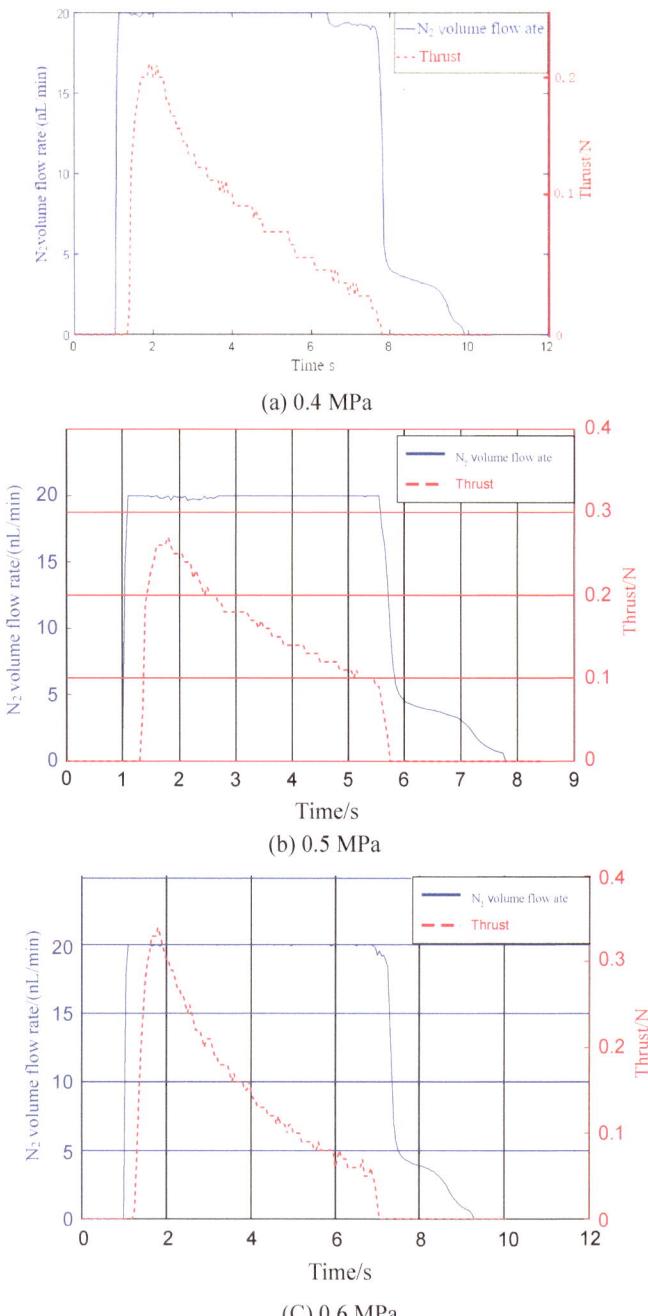

(a) 0.4 MPa

(b) 0.5 MPa

(C) 0.6 MPa

Fig. 10.1 The cold gas propulsion test results under different pressures

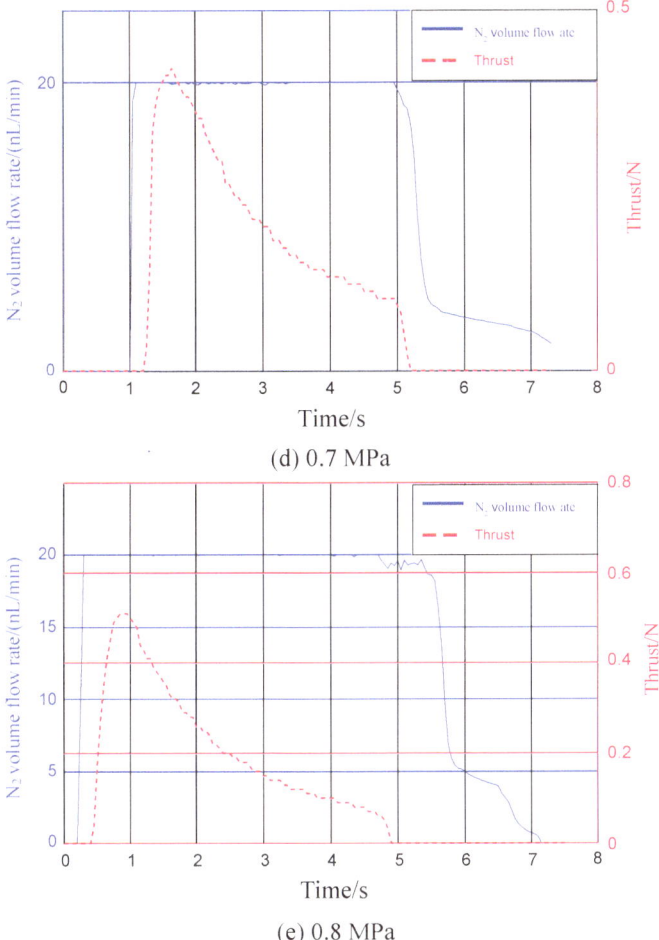

Fig. 10.1 (continued)

increases the resistance as propellant is ejected from the nozzle; simultaneously, the thrust sensor has limited accuracy, and the error in the measured value increases over time, so a thrust of zero can be experimentally obtained. Therefore, the maximum value of the thrust curve is taken as the measured nozzle thrust.

The thrust of cold gas propulsion under different pressures is shown in Table 10.1. The experimental data show that, when the gas cylinder pressure is fixed at 0.4 MPa, no matter how the flow rate is adjusted (the maximum is 20.00 nL/min), the maximum thrust is 0.2 N. Compared to the design conditions of 0.1, 0.2, 0.3, 0.4 and 0.5 N, the propellant flow rate is too high when the pressure is kept constant, and the flow rate needs to be appropriately reduced under each pressure to meet the desired thrust.

Table 10.1   Variation of thrust versus pressure

Pressure/MPa	0.4	0.5	0.6	0.7	0.8
Thrust/N	0.2	0.27	0.34	0.42	0.51
Desired thrust/N	0.1	0.2	0.3	0.4	0.5

With the pressure under each working condition being constant, the voltage of the flow controller is adjusted to achieve the desired thrust under each condition. The test results are shown in Fig. 10.2. At 20 °C and 101.325 kPa, the nitrogen density is 1.1654 kg/m^3, and the corresponding mass flow rates of nitrogen under each working condition are shown in Table 10.2.

Analysis shows that the specific impulse of the thruster in the cold gas test is low. At 0.4 MPa, the specific impulse is 48 s. However, at 0.4 MPa, the theoretical specific impulse value of N_2 at room temperature is approximately 75 s. The possible reasons

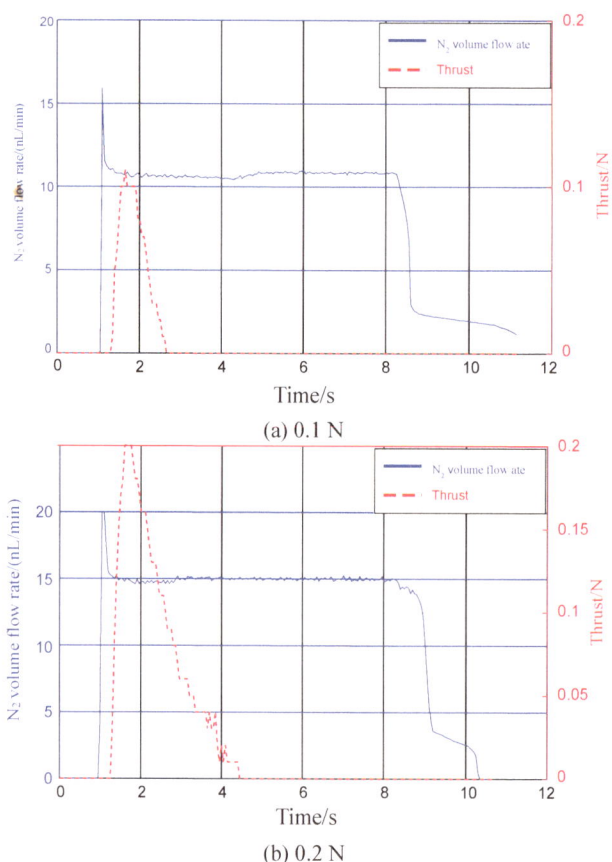

(a) 0.1 N

(b) 0.2 N

Fig. 10.2   Cold gas test curves under design conditions

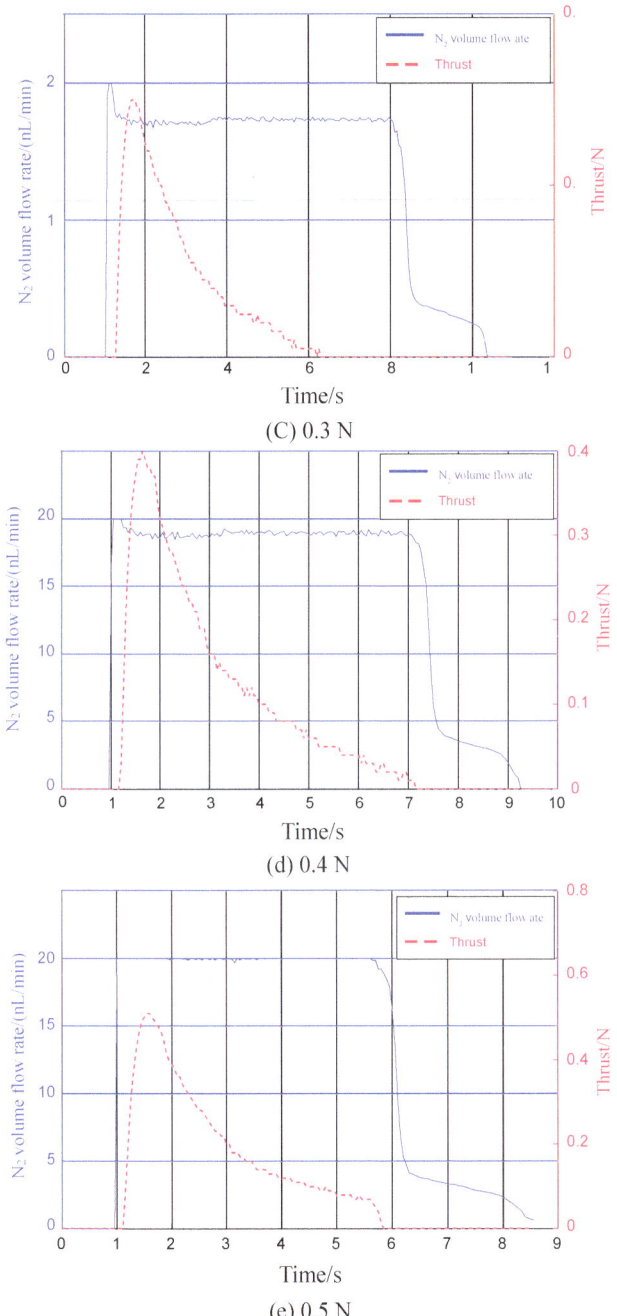

(C) 0.3 N

(d) 0.4 N

(e) 0.5 N

Fig. 10.2 (continued)

Table 10.2   Cold gas propulsion experiment results under design conditions

Pressure/MPa	0.4	0.5	0.6	0.7	0.8
Thrust/N	0.1	0.2	0.3	0.4	0.5
Flow rate/(nL/min)	10.74	15.01	17.26	19.01	20.00
Mass flow rate/($\times 10^4$ kg/s)	2.086	2.915	3.352	3.692	3.885
Specific impulse/s	48	68	89	108	128

for the low experimental value are as follows: the working fluid flow rate into the nozzle is low due to some leakage of the working fluid caused by the sealing issue during the installation of the thruster; the pressure in the vacuum chamber is still high after evacuation, and the experimental value is approximately 12.8 Pa; the thruster nozzle is designed to use hydrogen as a propellant, so there is a certain difference as nitrogen is used in the test.

The cold gas experiment clearly shows that the pressure of the gas cylinder has a great impact on the thrust. To study the effects of the flow rate and solar radiation power on the thrust, the pressure of the propellant cylinder is kept constant at 0.4 MPa.

First, the feasibility of heating by a xenon lamp light source is verified, then a thrust test is conducted on the thruster under low-power heating conditions, and the test result curves are shown in Fig. 10.3. The test results show that the flow controller has better stability, and the heating effect of the thruster is more obvious. The measured thrust increases from 0.04 N without heating to 0.08 N when heated by a 700 W xenon lamp and reaches 0.12 N when heated by a 2 kW xenon lamp.

10.3   Propellant Heating Experiment of Thruster Under Variable Working Conditions

Since the xenon lamp spectrum is basically consistent with the solar spectrum, the xenon lamp light source is used to simulate the heating of the thruster by sunlight. The variation range of the electric power of the xenon lamp is 700 W to 7 kW, and the electro-optic conversion efficiency is approximately 60%. In the experiment, the power is gradually increased to prevent the rupture of the secondary concentrator caused by the large thermal shock. Figure 10.4 shows the heating test in progress.

After the thruster is heated for a long time, the fixation panel of the concentrator burns and discolors to a certain extent, but the secondary concentrator remains intact, as shown in Fig. 10.5a, indicating that the secondary concentrator can work continuously for a long time in this high-temperature environment. The heat exchanger core material of the thruster is molybdenum metal. As shown in Fig. 10.5b, the structure is stable after heating. Therefore, the use of molybdenum metal as the material of the thruster can work for a long time in this harsh environment.

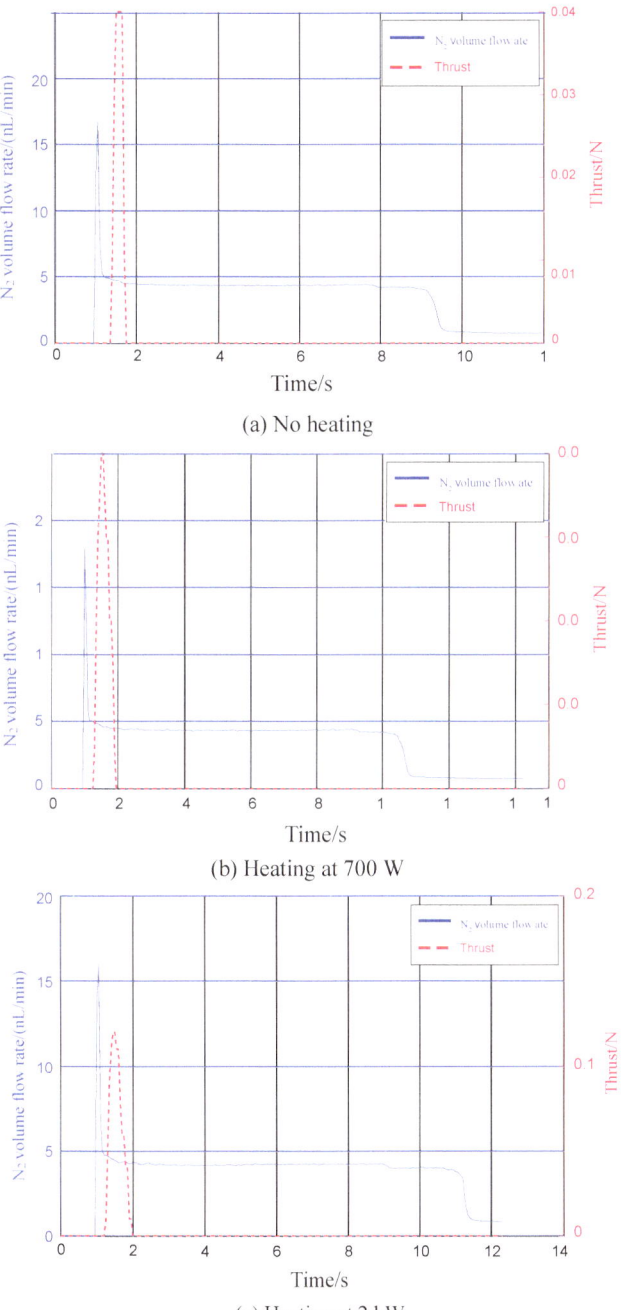

Fig. 10.3 Results of the low-power heating experiment

Fig. 10.4 Using a xenon lamp to heat the thruster

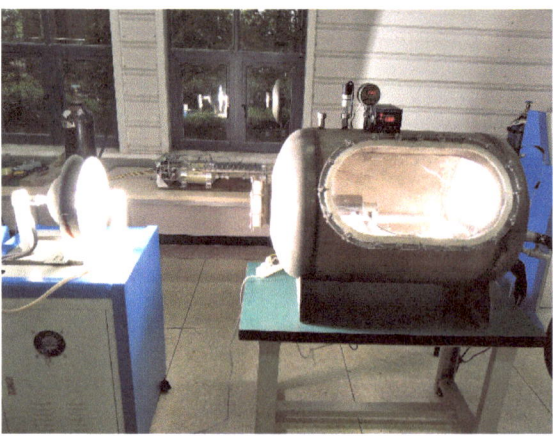

(a) Concentrator and fixation panel   (b) Thruster absorber

Fig. 10.5 Photographs of thrusters after heating

10.3.1 Adjusting the Flow Rate While Keeping the Heating Power Constant

To ensure the safety of the experiment, the pressure reducing valve is adjusted to maintain the pressure at 0.4 MPa, the heating power is 3 kW, and the voltage of the flow controller is adjusted to change the flow rate. Table 10.3 lists the flow rates measured when the flow controller is under different voltages.

Table 10.3 Working fluid flow rate under different voltage values of the flow controller

Voltage/(V)	1.0	2.0	3.0	4.0	5.0
Flow rate (nL/min)	4.20	8.40	12.24	16.31	20.00

The experimental thrust and flow rate results are shown in Fig. 10.6. The test results show that when the heating power remains constant, the thrust increases with the increase in the flow rate within a certain range. From an experimental point of view, the effectiveness of two ways of changing the thrust is theoretically verified.

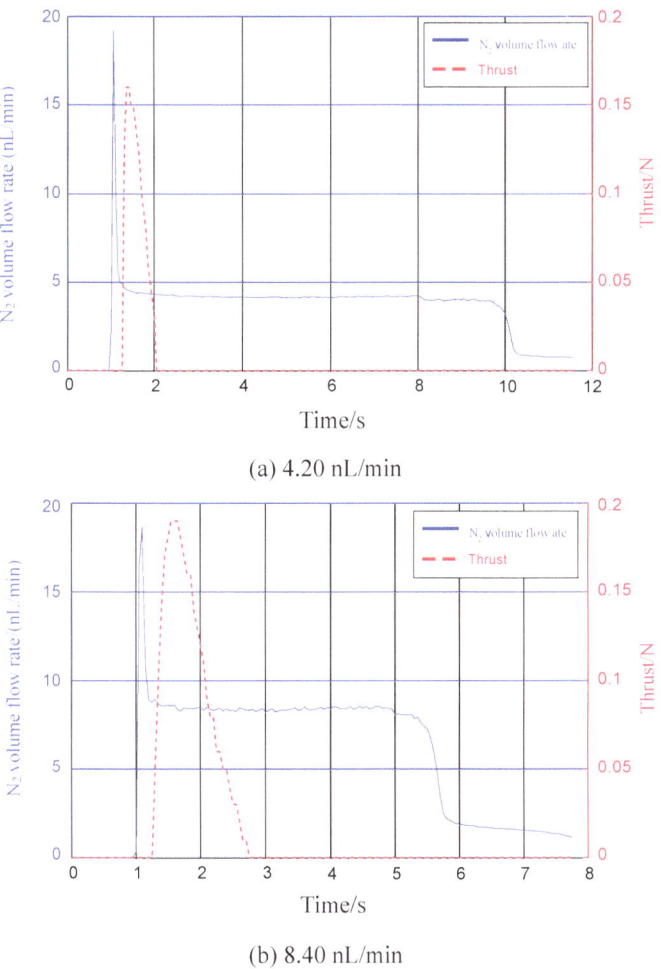

(a) 4.20 nL/min

(b) 8.40 nL/min

Fig. 10.6 Thrust variations under different flow rates

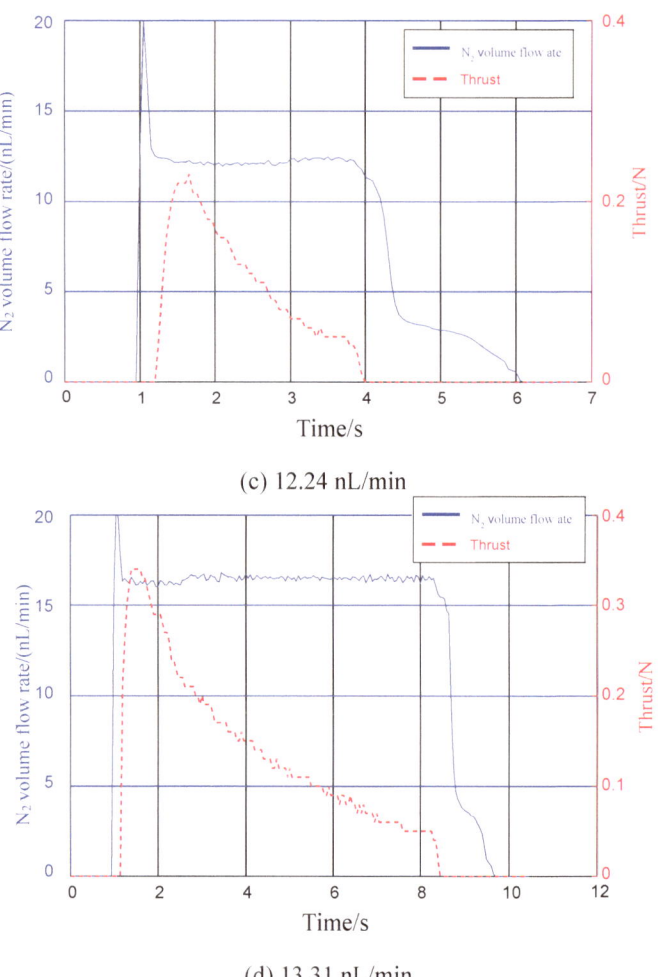

(c) 12.24 nL/min

(d) 13.31 nL/min

Fig. 10.6 (continued)

10.3.2 Adjusting the Heating Power While Keeping the Flow Rate Constant

The power of the xenon lamp to simulate the incident sunlight gradually increases, and electric powers of 1, 2, 3 and 4 kW are used to heat the thruster. The obtained test curves are shown in Fig. 10.7. Figure 10.7 shows that the maximum thrust of the thruster significantly increases with increasing heating power. Nitrogen is used as the propellant in the experiment, and the measured maximum thrust reaches 0.60 N.

In this chapter, experiments on solar thermal thruster under variable working conditions are carried out. During the cold gas propulsion experiments, due to the limitation of the structural parameters of the nozzle and the experimental conditions, the thruster could not operate at the design condition when only the propellant flow rate is adjusted. In the heating experiment, a xenon lamp is used to simulate sunlight, nitrogen is used as a propellant to verify the effectiveness of xenon lamp heating, the thruster thrust is significantly increased, and nitrogen can be heated to above 700 K. Therefore, from the experimental point of view, the feasibility of changing the thrust of the thruster by adjusting the flow rate and the incident sunlight is verified.

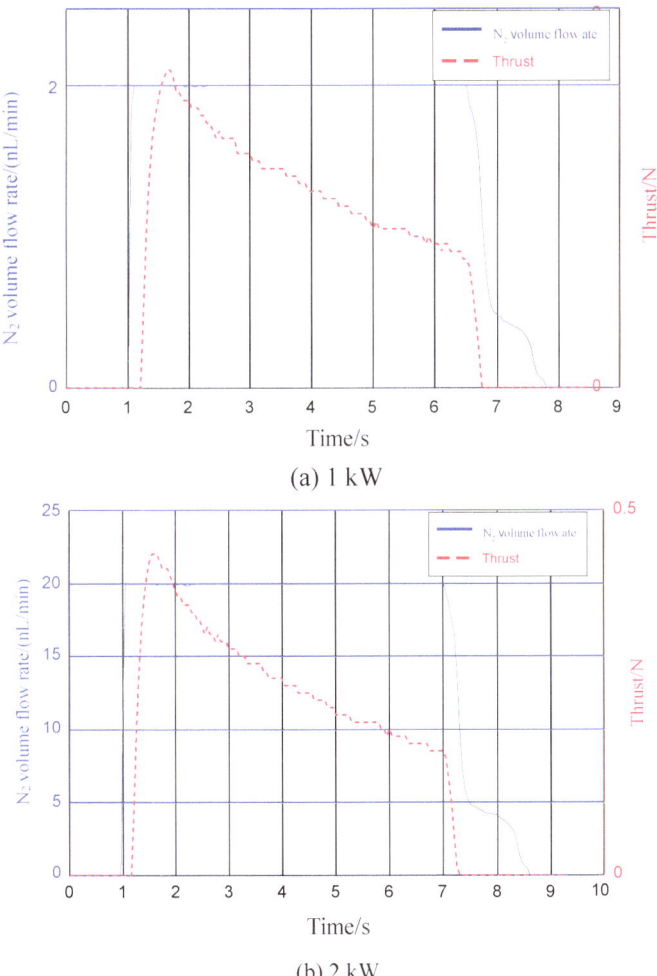

Fig. 10.7 Variations of flow rate and thrust under different xenon lamp powers

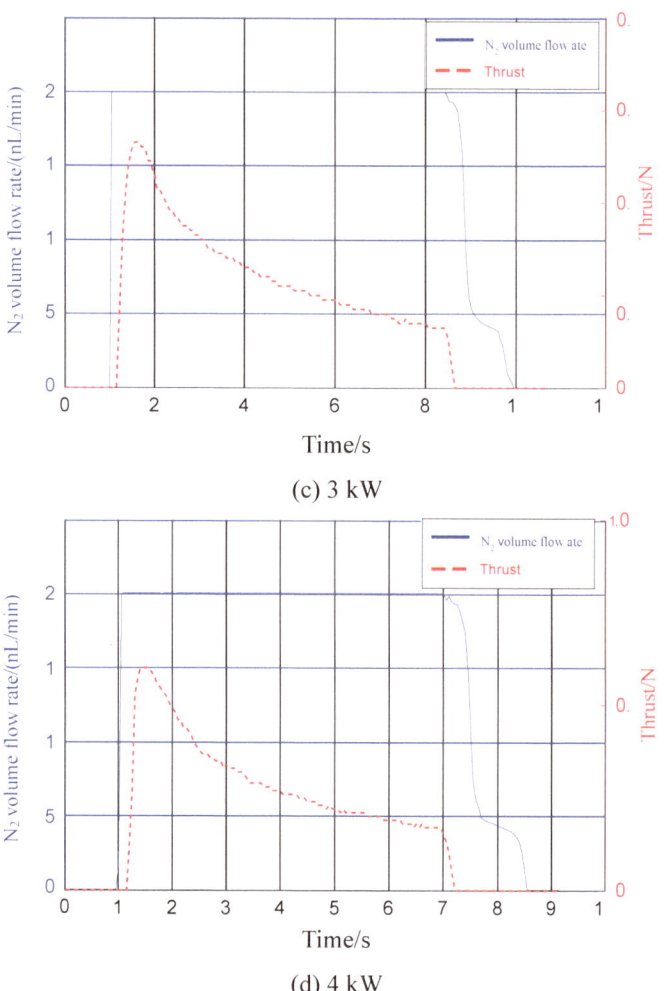

(c) 3 kW

(d) 4 kW

Fig. 10.7 (continued)

References

1. Kennedy FG, Palmer PL (2002) Preliminary design of a micro-scale solar thermal propulsion system. AIAA 2002-3928
2. Kennedy FG, Palmer PL (2002) Propulsion for space transportation of the 21st century. In: 6th international symposium, May 2002, Versailles, France
3. Wilner B, Hays L, Buhler R Research and development studies to determine feasibility of a solar LH2 rocket propulsion system. RTD-TDR-63-1085.
4. Morio S, Katsuya I, Yoshihiro N (2000) Very small solar thermal thruster made of single crystal tungsten for micro/nanosatellites. AIAA 2000-3832
5. Bérend N (2003) System study for a solar thermal thruster with thermal storage. In: 39th AIAA/ASME/SAE/ASEE joint propulsion conference and exhibition 20–23 July 2003, Huntsville, Alabama
6. Koroteev AS, Kochetkov YM, Akimov VN (2004) Solar power propulsion system adaptation to Ariane 5 and preliminary development plan. AIAA 2004-4173
7. Lyman RW, Ewing ME, Krishnan RS (2001) Solar thermal propulsion for an interstellar probe. In: 37th AIAA/ASME/SAE/ASEE joint propulsion conference and exhibition 8–11 July 2001, Salt Lake City, Utah
8. Zhang CL, Gao F, Zhang ZP et al (2004) Investigation and development on solar thermal propulsion. J Propul Technol 25(2):187–192
9. Xia GQ, Mao GW, Tang JL et al (2005) Research and development of solar thermal propulsion. J Solid Rocket Technol 28(1):10–14
10. Zhang CL, Zhang ZP, Wei ZM (2004) Design of concentrator for solar thermal propulsion. J Aerosp Power 19(4):557–561
11. Zhang CL, Wang P (2006) Flow and performance calculation of absorber/thruster of solar thermal propulsion. J Aerosp Power 21(5):943–948
12. Xia GQ, Mao GW, Tang JL et al (2005) Performances prediction of solar thermal propulsion system with refractive secondary concentrator. J Solid Rocket Technol 28(2):79–82
13. Huang MC, Du YL (2016) Air inlet property analysis of the air-breathing solar thermal propulsion. J Natl Univ Defense Technol 38(6):59–63
14. Valentian D, Amari M, Fratacci G (2002) A comparison of low cost cryogenic propulsion and solar thermal propulsion for orbit transfer. AIAA paper AIAA 2002-3590
15. Pearson JC, Lester DM, Holmes MR (2003) Solar thermal vacuum testing of an integrated membrane concentrator system at the NASA GRC Tank 6. AIAA 2003-5173
16. Clark P, Streckert H, Desplat JL (2003) Solar thermal test of cylindrical inverted thermionic converte. AIAA 2003-6102
17. Henshall P, Palmer P (2006) Solar thermal propulsion augmented with fiber optics: technology development. In: 42nd AIAA/ASME/SAE/ASEE joint propulsion conference and exhibit, July 2006, Sacramento, California

18. Partch R, Frye P (1999) Solar orbit transfer vehicle space experiment conception design. AIAA 99-2476
19. Olsen AD, Cady EC, Jenkins DS et al (1999) Solar thermal upper stage cryogen system engineering checkout test. In: 35th AIAA/ASME/SAE/ASEE joint propulsion conference, June 1999, Los Angeles, California
20. Etheridge F Solar rocket concept analysis: final technical report. AFRPL-TR-79-79
21. Humble R, Henry G, Larson W (1995) Space propulsion analysis and design
22. Schleiniztz J, Lo R Solar thermal OTVs in comparison with electrical and chemical propulsion systems. In: 38th congress of the international astronautical federation. IAF-87-199, Brighton, UK
23. Tucker S (2001) Solar thermal engine testing. In: NASA JPL/MSFC/UAH 12th annual advanced space propulsion workshop, April 2001, Huntsville, Alabama
24. Patrick EF, Kudija CT (1998) Integrated solar upper stage engine ground demonstration test result and data analysis. In: 34th AIAA joint propulsion conference and exhibit, July 1998, Cleveland, Ohio
25. Clark W, Alan M (1995) Conceptual design of a solar thermal upper stage (ISUS) flight experiment. AIAA 95-2842
26. Woodcock G, Byers D (2003) Results of evaluation of solar thermal propulsion. AIAA2003-5029
27. Richard F, Gregory T Design description of the ISUS receiver/absorber/converter configuration and electrical test. AIAA96-3046
28. Brian A (1996) Analysis of the solar thermal upper stage technology demonstrator liquid acquisition device with integrated thermodynamic vent system. AIAA96-2745
29. Nakamura T, Sullivan D, McClanahan JA et al (2004) Solar thermal propulsion for small spacecraft. AIAA 2004-4138
30. Nakamura T, Krech RH, McClanahan JA et al (2005) Solar thermal propulsion for small spacecraft: engineering system development and evaluation. AIAA 2005-3923
31. Sahara H (2001) Opposed-cavity solar thermal thruster made of single crystal tungsten. In: 2001 international electric propulsion conference, 27th IEPC, Pasadena, CA, USA
32. Sahara H, Shimizu M (2004) Solar thermal propulsion system for microsatellite orbit transferring. AIAA 2004-3764
33. Shimizu M, Naito H, Sahara H (2001) 50mm cavity diameter solar thermal thruster made of single crystal molybdenum. AIAA 2001-3733
34. Sahara H, Shimizu M (2003) Solar thermal propulsion investigation activities in NAL. In: Komurasaki K (ed) Proceedings of the second international symposium on beamed energy propulsion, AIP conference proceedings, 702, Sendai, Japan, pp 322–333
35. Sahara H, Shimizu M (2003) Solar thermal propulsion system for a Japanese 50 kg-class microsatellite. AIAA 2003-5032
36. Li X, Fan G, Zhang Y et al (2018) A Fresnel concentrator with fiber-optic bundle based space solar power satellite design. Acta Astronaut 153:122–129
37. Zhou R, Wang R, Xing C et al (2022) Design and analysis of a compact solar concentrator tracking via the refraction of the rotating prism. Energy 251:125–137
38. Fan G, Duan B, Zhang Y et al (2021) Secondary concentrator design of an updated space solar power satellite with a spherical concentrator. Sol Energy 214(5):400–408
39. Laug K (1989) The solar propulsion concept is alive and well at the astronautics laboratory. Chem Propul Inf Agency 515:267–285
40. Miles BJ, Kerr JM (1996) Coatings for high temperature graphite thermal energy storage in the integrated solar upper stage. AIAA 96-3047
41. Kessler TL (2001) An overview of a solar thermal propulsion and power system demonstration applicable to HEDS[R]. AIAA 2001-4777
42. Tucker TW, Landrum DB (1996) Effects of scale and chamber conditions on the performance of hydrogen thrusters for solar thermal rockets. AIAA 96-3216
43. Stark LD, Bonometti JA, Gregory DA (1996) Experimental evaluation of solar thermal rocket absorber and concentrator surfaces. AIAA96-2929

44. Henshall PR (2006) A proposal to develop and test a fibre-optic coupled solar thermal propulsion system for microsatellites. Approved for Public Release
45. Duffie JA, Beckman WA (1980) Solar engineering of thermal processes. China Science Publishing & Media Ltd., Beijing
46. Siegel R, Howell JR (1990) Thermal radiation heat transfer. China Science Publishing & Media Ltd., Beijing
47. Shi YC (2009) Solar energy principles and technology. Xi'an: Xi'an Jiaotong University Press Co., Ltd
48. Jonathan AS, George DQ (2009) Failure analysis of sapphire refractive secondary concentrators. NASA/TM-2009-215802
49. Dai GL, Xia XL, Sun C (2011) Concept design and performance analysis for TPV system of solar thermal propulsion. J Astronaut 32(2):451–457
50. Dai GL, Xia XL, Sun C (2011) The transport characteristics of concentrated solar radiation in volumetric receiver. J Eng Thermophys 32(11):1941–1944
51. Dai GL, Xia XL, Yu MY (2010) Performance calculation and analysis of multi-STP. J Astronaut 31(6):1631–1636
52. Dai GL (2012) Optical heat transfer characteristics of two-stage solar energy concentration and high-temperature thermal conversion. Harbin Institute of Technology
53. Dai GL, Xia XL (2010) Thermal conversion characteristics of solar energy receiver with quartz window. J Eng Thermophys 31(6):1005–1008
54. Xia GQ (2005) Solar thermal conversion mechanism research and experimental system design of STP. Northwestern Polytechnical University, Xi'an
55. Yang J, Yang LJ (2010) Effects of propellant heat groove configuration on solar thermal propulsion performance. J Aerosp Power 25(5):1156–1162
56. Shoji JM, Frye PE, McClanahan JA (1992) Solar thermal propulsion status and future. AIAA92-1719
57. Su XY, Zhang BZ, Fang YJ et al (2004) Research on the characteristics of fluid flow and heat transfer in a rotating helical rectangle duct. Chin J Appl Mech 21(4):46–50
58. Chen HJ (2003) Flow structure and heat transfer characteristics in rotating curved pipes. Zhejiang University, Hangzhou
59. Ma JF (2007) Turbulent flow structure and heat transfer characteristics in a rotating curved pipe. Zhejiang University, Hangzhou
60. Robert L (2006) Overview of United States space propulsion technology and associated space transportation systems. J Propul Power 122(6):1310–1333
61. White SM (2010) High-temperature spectrometer for thermal protection system radiation measurements. J Spacecr Rocket 47(1):21–28
62. Yu XH (2001) Study of cooling characteristics of laminates. Northwestern Polytechnical University, Xi'an
63. Yu XH, Dong ZR, Liu SL et al (2000) Study for flow resistance characteristics of the modelled laminated porous wall. J Propul Technol 21(4):47–50
64. Yu XH, Quan DL, Xu DC et al (2003) Investigation of the internal heat transfer characteristics of lamilloy. Acta Aeronautica ET Astronautica Sinica 24(5):405–410
65. Quan DL (2005) Study of cooling characteristics of laminate blades. Northwestern Polytechnical University, Xi'an
66. Quan DL, Yu XH, Liu SL et al (2003) Experimental and numerical investigation of internal-flow resistance characteristics in laminate porous plates. J Propul Technol 24(5):425–428
67. Quan DL, Li JH, Liu SL (2004) Numerical investigation of cooling characteristics in lamilloy with snowflake design. J Therm Sci Technol 3(1):55–59
68. Niu L (2002) Research on platelet regenerative cooling technology for liquid rocket motor. Shanghai Jiao Tong University, Shanghai
69. Yang WH, Cheng HE, Wang PY et al (2004) Structural design analysis for platelet transpiration cooling section at throat of thrust chamber. J Propul Technol 25(4):316–319
70. Yang WH, Cheng H, Cai A (2002) Experimental study for flow characteristics of adjustment channel in platelet thrust chamber. AIAA Paper 2002-3705

71. Liu WQ, Chen QZ, Wu BY (1998) Calculation method of heat transfer in platelet transpiration cooled thrust with typical structure. J Propul Technol 19(6):15–19
72. Liu WQ, Chen QZ The effect of transpiration cooling with liquid oxygen on the flow field. AIAA Paper 98-3515
73. Zhang F (2008) Theoretical analysis and applied research on transpiration cooling of laminates. National University of Defense Technolog, Changsha
74. Karniadakis GE, Beskok A (2002) Micro flows: fundamentals and simulation. Springer Verlag, New York
75. Pfahler JN, Harley JC, Bau H et al (1991) Gas and liquid flow in small channels. ASME DSC 32:49–60
76. Harley JC, Huang YF, Ban H et al (1995) Gas flow in microchannels. Fluid Mech 284:257–274
77. Arkilic E, Breuer K (1993) Gaseous flow in small channels. AIAA 93-3270
78. Arkilic E, Breuer K, Schmidt M (1994) Gaseous flow in microchannels. Appl Microfabr Fluid Mech 197:57–66
79. Jiang XN (1996) Experimental study of microfluidic measurement and control system. Tsinghua University, Beijing
80. Qin FH, Yao JC, Sun DJ (2001) Experimental measurement for the gas flow rates in microscale circular pipe. J Exp Mech 16(2):119–126
81. Guo ZY (2000) Frontier of heat transfer-microscale heat transfer. Adv Mech 30(1):1–6
82. Wu XB, Guo ZY (1997) Characterization of gas flow and heat transfer in microfine smooth tubes. J Eng Thermophys 18(3):326–330
83. Du DX, Li ZX, Guo ZY (1999) Effect of reversible work and viscous dissipation on gas flow characteristics in a microtube. J Tsinghua Univ 39(11):58–60
84. Cai CP, Boyd ID, Fan J (1999) Direct simulation methods for low-speed microchannel flows. AIAA 99-3801
85. Fan Q, Shen Q (2002) Micro-scale gas flows. Adv Mech 32(3):321–336
86. Carlson HA, Roveda R, Boyd ID A hybrid CFD-DSMC method of modeling continuum-rarefied flows. AIAA 2004-1180
87. Qi ZG (2007) Direct simulation Monte Carlo on micro-and nanoscale gas flow and heat transfer. University of Chinese Academy of Sciences, Beijing
88. Zheng L (2010) Lattice Boltzmann method for microscale flow and heat and mass transfer. Huazhong University of Science and Technology, Wuhan
89. Lacy JM, Carmack WJ, Miller BG (2001) Results of risk reducation activities for the SOTV space experiment solar engine. AIAA 2001-3989
90. Kudija C (1996) The integrated solar upper stage (ISUS) engine ground demonstrator (EGD). AIAA 96-3043
91. Cady EC, Olsen Jr AD (1996) Cryogen storage and propellant feed system for the integrated solar upper stage (ISUS) program. AIAA 96-3044
92. Richards DR, Vonderwell DJ (1997) Flow network analyses of cryogenic hydrogen propellant storage and feed system. AIAA97-3223
93. Tucker TW, Landrum DB (1996) Effects of scale and chamber conditions on the performance of hydrogen thrusters for solar thermal rockets. AIAA Paper 96-3216
94. Landrum DB, Beard RM (1996) Dual fuel solar thermal propulsion computational assessment of nozzle performance. AIAA Paper 96-3217
95. LeBar JF (1997) Testing of multiple orifice Joule-Thomson devices in liquid hydrogen. AIAA97-3315
96. Davidson DF, Kohse-Hoingaus K, Chang Y (1990) A pyrolysis mechanism for ammonia. Int J Chem Kinet 22:513–535
97. Konnov AA, Ruyck JD (2000) Kinetic modeling of the thermal decomposition of ammonia. Combust Sci Technol 152:23–37
98. Chambers A, Yoshii Y, Inada T (1996) Ammonia decomposition in coal gasification atmospheres. Can J Chem Eng 74:929–934
99. Monnery WD, Hawboldt KA, Pollock AE (2001) Ammonia pyrolysis and oxidation in the Claus furnace. Ind Eng Chem Res 40:144–151

100. Darcy LA, Pavlos GM (2004) Pulsed inductive thruster, part 2: two-temperature thermochemical model for ammonia. AIAA 2004-4092
101. Colonna G, Capitta G, Capitelli M et al. (2006) Model for ammonia solar thermal thruster. J Thermophys Heat Transfer 20(4):772–779
102. Dagmar B, Monika AK, Helmut LK (2006) Design of an ammonia propellant feed system for a 1 kW class thermal arcjet thruster system. AIAA 2006-4852
103. Zhang GJ, Cai XD, Liu MH et al (2004) Cold flow field and propulsive performance of a micro Laval nozzle. J Propul Technol 25(1):54–57
104. Zhang YH (2002) Thermal analysis of space solar thermal absorber and storage. Beihang University, Beijing
105. Wu LJ (2005) Optimization of blast furnace cooling wall structure and intelligent simulation method based on heat transfer analysis. Shanghai Jiao Tong University, Shanghai
106. Tao WQ (2001) Numerical heat transfer, 2nd edn. Xi'an Jiaotong University Press Co., Ltd., Xi'an
107. Yu JZ (2006) Heat exchanger principle and design. Beijing University Publishing House, Beijing
108. Yu QZ (2000) Principle of radiant heat transfer. Harbin Institute of Technology Publishing House, Harbin
109. Liu LH, Zhao JM, Tan HP (2008) Finite element and spectral element methods for numerical simulation of the radiation transfer equation. China Science Publishing & Media Ltd., Beijing
110. Tan HP, Xia XL et al (2006) Numerical calculations of infrared radiation characteristics and transmission—computational thermal radiative transfer. Harbin Institute of Technology Publishing House, Harbin
111. Fan XJ (2004) Pneumatic heating and thermal protection systems. China Science Publishing & Media Ltd., Beijing
112. Li ZX, Guo ZY (2010) Field synergy theory for convective heat transfer optimization. China Science Publishing & Media Ltd., Beijing
113. Zhu D, Nathan S J, Miller RA (2009) Thermal-mechanical stability of single crystal oxide refractive concentrators for high-temperature solar thermal propulsion. NASA/TM—1999-208899
114. Kennedy FG (2004) Solar thermal propulsion for microsatellite maneuvering. University of Surrey, Guildford
115. Shi YY, Na HY (2009) Design, preparation and evaluation of solar spectrally selective absorption films. Tsinghua University Press, Beijing
116. Zhang HF (2007) Principles of solar thermal utilization and computer simulation, 2nd edn. Northwestern Polytechnical University Press Co. Ltd., Xi'an
117. Kitamura R, Pilon L, Jonasz M (2007) Optical constants of silica glass from extreme ultraviolet to far infrared at near room temperature. Appl Opt 46(33):8118–8133
118. Beder EC, Bass CD, Shackleford WL (1971) Transmissivity and absorption of fused quartz between 0.22 and 3.5 from room temperature to 1500°C. Appl Opt 10(10):2263–2267